SHAPING SUSTAINABLE FASHION

changing the way we make and use clothes

EDITED BY ALISON GWILT & TIMO RISSANEN

publishing for a sustainable future

LONDON · WASHINGTON, DC

First published in 2011 by Earthscan

Earthscan Ltd, Dunstan House, 14a St Cross Street, London EC1N 8XA, UK
Earthscan LLC, 1616 P Street, NW, Washington, DC 20036, USA

Earthscan publishes in association with the International Institute for Environment and Development

For more information on Earthscan publications, see **www.earthscan.co.uk** or write to **earthinfo@earthscan.co.uk**

ISBN: 978-1-84971-241-5 Hardback
      978-1-84971-242-2 Paperback

Typeset by Amy Common | Sai Designs **www.saidesigns.com.au**
Cover design by Rogue Four Design **www.roguefour.com**
Cover illustration by Amy Common | Sai Designs **www.saidesigns.com.au**

A catalogue record for this book is available from the British Library

Library of Congress Cataloging-in-Publication Data
Gwilt, Alison.
  Shaping sustainable fashion : changing the way we make and use clothes / Alison Gwilt and Timo Rissanen.
    p. cm.
  Includes bibliographical references and index.
  ISBN 978-1-84971-241-5 (hardback) -- ISBN 978-1-84971-242-2 (pbk.) 1. Fashion design. 2. Textile design. 3. Sustainable design. I. Rissanen, Timo. II. Title.
  TT507.G95 2010
  746.9'2--dc22

                    2010032968

At Earthscan we strive to minimize our environmental impacts and carbon footprint through reducing waste, recycling and offsetting our CO_2 emissions, including those created through publication of this book. For more details of our environmental policy, see **www.earthscan.co.uk**.

This book was printed in the United Kingdom by Butler, Tanner & Dennis, an ISO 14001 accredited company. The paper used is certified by the Forest Stewardship Council (FSC), and the inks are vegetable based.

In Memory of Ronald William Mason

CONTENTS

4

LIST OF FIGURES

ACKNOWLEDGEMENTS

We wish to thank our authors Joan Farrer, Marie O'Mahony, Holly McQuillan, Kathleen Dombek-Keith, Suzanne Loker, Jana Hawley and Kate Fletcher for sharing their valuable insight and knowledge. We would also like to thank the following designers and researchers who have contributed to the case studies; Luke Sales and Anna Plunkett at Romance Was Born and Rae Begley at Little Hero; Jennifer Shellard; Bijan Sheikhlary; Mike Betts and Mark Holt at Better thinking Ltd; Dr Gene Sherman, Executive Director of the Sherman Contemporary Art Foundation; Leonie Jones at the Powerhouse Museum; Alex Martin; Delores D'Costa at The Smith Family; Helen Storey. In addition we would like to thank the many designers, researchers and companies who supplied us with images of their work.

We also wish to thank: Michael Fell, Claire Lamont and Anna Rice at Earthscan for their support; our production team in Sydney, Amy Common and Deborah Turnbull for their commitment and enthusiasm for the entire Fashioning Now project; Tania Creighton and Holly Williams, curators at the UTS gallery in Sydney, Jonathan James for exhibition design in Sydney and Jasmin Stephens at the Fremantle Arts Centre, for staging the Fashioning Now exhibitions; staff at the School of Design at the University of Technology Sydney and Parsons The New School for Design.

Finally, we would like to thank Ian and Dylan, and George for their continued love and support.

This project has been assisted by the New South Wales Government through its Environmental Trust.

LIST OF CONTRIBUTING AUTHORS

JOAN FARRER has 32 years commercial fashion, textiles, branding and research and development expertise with international industrial retailers, institutions and non-governmental organizations (NGOs) in design, manufacture and policy development. Her vanguard Royal College of Art (RCA) PhD in 2000 concentrated on global supply chain analysis, discussing economic, social and environmental production (sustainability). Current research is focused on Sustainable and Smart solutions with transdisciplinary collaborators. Farrer's academic roles include Director MA Fashion, Senior Research Fellow, Director Research Lab, Associate Professor and Reader.

MARIE O'MAHONY is Professor of Advanced Textiles for Fashion Design at University of Technology, Sydney (UTS) and Visiting Professor at University of the Arts, London. Her role at UTS is divided between research and teaching, the latter including setting up a new Masters course. She has worked for a number of years as a consultant with clients including Hussein Chalayan, and curated exhibitions for the Stedelijk Museum and British Council amongst others. She has written and co-authored a number of books including the TechnoTextiles series and is currently working on a new book to be published with Thames and Hudson in 2011. She is a member of the Australian Government's Textile, Clothing and Footwear Industries Innovation Council (TCFIIC).

HOLLY MCQUILLAN is a lecturer in design at Massey University, College of Creative Arts in Wellington, New Zealand with a BDes and MDes in Fashion Design. Her research focuses on sustainable design practice within a contemporary material culture framework with a particular interest in production and consumption systems within the fashion world. Using innovative drape and garment creation techniques she explores new ways of making and consuming clothes that have been exhibited in galleries internationally.

KATHLEEN DOMBEK-KEITH earned an MA degree in apparel design at Cornell University, and her master's thesis, 'Re-Fashioning the Future: Eco-Friendly Apparel Design', was recently published. Her research focuses on reducing the environmental impacts of clothing, specifically laundering and materials, and fostering meaningful relationships between wearers and their clothing through innovative design approaches. She is currently a lecturer in the Department of Apparel Merchandising and Interior Design at Indiana University, Bloomington.

SUZANNE LOKER is a Professor Emerita in the Department of Fiber Science and Apparel Design at Cornell University. She has published widely on innovative business strategies in the apparel industry, specifically those involving socially responsible practices, and body scanner and mass customization technologies. She recently co-authored the book, *Social Responsibility in the Global Apparel Industry*, with Drs Marsha Dickson and Molly Eckman. She earned her BS and MA degrees in apparel design at the University of Wisconsin-Madison and Syracuse University, respectively, and her PhD degree in educational psychology at Kansas State University.

JANA HAWLEY is Professor and Department Chair of Textile and Apparel Management at the University of Missouri. She is currently President of the International Textile and Apparel Association. Hawley's scholarship focuses on textile recycling but she has also published in the areas of e-commerce and the Old Order Amish. Hawley is a Fulbright Scholar to India and a Global Scholar to Thailand. Her work in sustainability has gained her international recognition including work in Greece, India, Italy and South Korea. She serves on the Board for the Council of Textile Recycling.

KATE FLETCHER is a sustainable designer, consultant, writer and key opinion leader in fashion, textiles and sustainability. Her work – in academia, with high street retailers and NGOs – has been at the forefront of design for sustainability in fashion and textiles for the last 15 years. It has roots in ingenuity, vitality, care and resourcefulness and is fed by design ideas and practical action. Kate holds a PhD from Chelsea College of Art and Design, is Reader in Sustainable Fashion at London College of Fashion and the author of *Sustainable Fashion and Textiles: Design Journeys*.

9

FOREWORD

Education is a key component of the New South Wales (NSW) Government's commitment to foster sustainable development. Education is a vital tool because it helps people to understand the nature and complexity of environmental challenges and builds their capacity to take appropriate action. The NSW Government's Learning for Sustainability 2007–2010 plan calls on all sectors to play their part in building a learning society, one in which we are all informed and active contributors to creating a more sustainable future. Within the fashion industry, *Shaping Sustainable Fashion: Changing the Way We Make and Use Clothes* has achieved this by teaching us how we can match sustainability and fashion in a unique, contemporary and diverse way.

The Environmental Trust, an independent statutory body established by the NSW government, funds projects like the University of Technology Sydney's *Fashioning Now* project, from which this book originates. Supported by the NSW Environmental Trust's Environmental Education programme, Fashioning Now has delivered an array of enlightening learning methods, including sustainability workshops, symposiums and exhibitions featuring innovative research projects from Australian and international practitioners that investigate fashion and sustainability. All of this knowledge has been pulled together in this publication to disseminate the vast range of sustainable solutions currently being explored by designers, researchers and manufacturers.

This book is certainly a step forward in changing the face of a fashion industry that is characterized by a high level of waste among manufacturers and driven by a fast-paced cycle of seasonal products. *Shaping Sustainable Fashion* raises awareness of the problem of textile waste, and gives consumers and designers the opportunity to learn how to work sustainably through solutions on waste avoidance, waste management and resource recovery. Additionally, by holding both consumers and designers accountable, *Shaping Sustainable Fashion* demonstrates the importance of making informed choices for the environment.

Shaping Sustainable Fashion's lessons on how to 'source, make, use and last' should be explored and shared by designers, researchers and consumers alike to make for a more sustainable and ethical fashion industry.

— *Amy Rosser*

The Environmental Trust
Department for Environment, Climate Change and Water NSW

INTRODUCTION FROM THE EDITORS
Alison Gwilt & Timo Rissanen

Shaping Sustainable Fashion explores the issues of fashion, sustainability and specifically the way in which fashion clothing is produced, used and discarded. Today the fashion industry relies on the fast and efficient manufacture of new seasonal trend-driven products for an identified consumer in a competitive marketplace. The continued cycle of buying, using and disposing of fashion clothing is based upon a system of production that has serious consequences for our society and the environment. The trend for fast fashion has generated an exponential rise in the sale of fashion garments that are often worn too little, washed too often and quickly become discarded, with an estimated 30kg textile waste per person reaching UK landfill each year (Allwood et al, 2006). Moreover, while this trend is typically associated with garments available in the high street, it seems that at all levels of the fashion industry there continues to be a focus on the production of market driven and disposable goods. The intention of this publication is to bring to light a multiplicity of sustainable strategies that are being employed to reduce the textile waste generated during the manufacture and use of fashion clothing.

Within the book we present a diverse set of approaches to sustainability that are currently being discussed within design practice, yet at the same time we have embraced ideas that present promising scenarios for a future fashion industry. These discussions are intended to generate dialogue and debate. In fact our aim for the book is not to provide definitive answers but to question current methods of design practice and offer alternative scenarios that challenge how garments are produced and used. Fashion is often perceived negatively in terms of sustainability and yet one of its inherent qualities is innovation and the search for new solutions. We aim to reveal the various innovative ways in which fashion designers, makers and users are refashioning fashion for a sustainable future.

Although there is an increasing universal awareness of environmentalism and ethical issues, we recognise that the field of sustainable fashion can appear complex. Fashion designers and consumers are often confused by the language of sustainability and professional resources sometimes do not make it clear how people can connect with methods of best practice, creating barriers for engagement with sustainability. In light of this problem the book follows a lifecycle approach, from a design perspective, and divides academic papers and case studies into four distinct sections: *Source, Make, Use* and *Last*.

Source takes a look at the complexities associated with the sourcing and manufacturing of sustainable fashion materials and products in a globalized industry. This section explores the environmental and social impacts associated with the production of materials and garments, and in particular the significant affect that products and processes have on our natural resources. Crucially, the chapter reiterates the need for designers and consumers to make informed choices. The *Make* section focuses on the production of fashion clothing and in particular the role and influence that the fashion designer can have in changing the current practices applied within the manufacturing process. *Make* surveys the production methods applied in a number of sectors of the fashion industry, and highlights positive approaches and sustainable strategies that can be applicable and adaptable to other levels of the industry. Within the *Use* section we discuss the positive contribution that the consumer can have on the lifecycle of a fashion garment; in particular we explore the schemes that can engage the consumer to slow fashion consumption. Highlighting the way in which the consumer engages with fashion, through the selection, use, washing, care, repair and disposal of a garment remains as critical as those contributions made by the producers of fashion. Finally, in the section entitled *Last* we investigate alternative systems and approaches that may reduce the amount of clothing contributing to landfill waste. We reveal the system of textile recycling and of the reuse of waste materials into new products, and we suggest other diverse strategies that challenge the speed and significance of fashion in a changing technological and cultural landscape.

Some of the sustainable strategies that are being explored within the sections of this book range from slow fashion, product/service systems, designing for waste minimization and end-of-life strategies. We have attempted to illustrate the extensive range of possibilities available in sustainable fashion practice in addition to established thinking about fashion that is produced with organic or recycled materials. Moreover, we have been keen to reveal fascinating examples that may not fit traditional notions of sustainable fashion but nevertheless make a strong case for new, improved practices.

For the case study components of the publication, we draw on research projects from international creators and researchers who were engaged in an Australian fashion project, *Fashioning Now* (2009), which comprised an exhibition, industry symposium and a project website. The symposium drew attendees from across Australia and New Zealand, from educational institutes, government and NGOs, and the fashion industry. This suggested that there is a significant thirst for knowledge about environmental and social issues in relation to the production and consumption of fashion clothing across a broad spectrum of people. The *Fashioning Now* project received assistance from the New South Wales Government through its Environmental Trust, and it has kindly provided the Foreword for this text.

Shaping Sustainable Fashion intends to demonstrate to fashion and textile design students, fashion designers in industry and fashion consumers that through the act of designing and the use of responsible patterns of consumption, textile waste can be avoided and reduced. The fashion industry needs to positively respond to the view that developing garments at the best possible price is not the only way to conduct business, especially when there continues to be a growth in public interest for environmentally friendly and ethically produced goods. Our vision for the future of fashion is one where all fashion is considered sustainable, making it entirely unnecessary to label it so. And in so doing, reaching a point where society perceives fashion design and production as an inherently positive facet of our culture.

Allwood, J. M., Laursen, S. E., Malvido de Rodriguez, C. and Bocken, N. M. P. (2006) *Well Dressed? The Present and Future Sustainability of Clothing and Textiles in the United Kingdom*. Institute for Manufacturing, University of Cambridge, Cambridge

RIGHT | Detail, Wonderland twirling flowers. Photographer Alex Maguire

SOURCE
Chapter 1

SOURCE | INTRODUCTION

This chapter explores the complexities associated with the sourcing and manufacturing of sustainable fashion materials and products in a globalized industry. Sourcing within the fashion supply chain is typically associated with business: budgets and deadlines, purchasing and selling, shipping and supplying. This is complicated further with the interdependent relationships between supplier and designer, designer and maker, maker and seller, seller and user, all of whom are connected to a product that is often determined by price, quality and speed.

Designers play a significant role in the development of new fashion products and they can lead the selection of materials and services used within the production process. A designer may need to locate fabric suppliers, trimmings suppliers, textile dyers and finishers, manufacturers for sample runs, and so on. But what are the ethical and environmental dilemmas that arise as a consequence of this decision-making? Most designers or product developers would probably admit that they do not question the production processes involved in developing a fabric or recognize what negative environmental and social impacts may be associated with a fabric during the manufacture, use and disposal of a garment. This unquestioning approach is typical and perhaps understandable since a company's interest in or time available for research into sustainable fibres, materials and processes may be minimal. This then raises the question of how a designer should select materials and choose services.

Building a relationship with a responsible and well-informed supplier can alleviate some of these concerns. The knowledge shared through this trusted association can be pivotal in assisting the designer. As suppliers communicate the sustainability credentials of a fabric, or a service, the designer becomes empowered through knowledge. Undeniably designers need to better familiarize themselves with the materials and processes that they use and promote in the production of fashion. Fashion fabrics go through a number of production processes from growing or manufacturing fibres and yarns, through to the dyeing and processing of fibres into fabrics. However, few designers would recognize the negative impacts of a fabric that are felt through the entire lifecycle of the garment, beyond fibre and textile production through to garment manufacture and disposal. Coming to terms with this fact will provide the designer the opportunity to reduce negative impacts whilst at the same time maximizing positive impacts; this should be the fundamental goal.

Joan Farrer ponders the idea of achieving sustainability in a global fashion and textiles industry as a utopian ideal. Farrer discusses the significance of the sector in relation to people, while highlighting the physical impact of the industry on the environment. By drawing together solutions for a more sustainable fashion industry Farrer also raises the need for a change in mindset by consumers, manufacturer and retailers. Meanwhile, Marie O'Mahony explores the issue of materials and the argument relating to water usage in the production and care of natural and man-made fibres. In comparing the benefits and drawbacks of both types of fibres and materials O'Mahony poses the thought that the future of materials may lie in hybrid combinations, particularly if water becomes increasingly short in supply.

JOAN FARRER

1.1 REMEDIATION:
Discussing Fashion Textiles Sustainability

Remediation: (noun) acting as a remedy or solution to a problem; in this case the use of remedial methods to improve learning skills to reverse social and environmental damage.

INTRODUCTION

This essay will review the idea of fashion (and by association textiles) sustainability to establish if it is a utopian ideal by looking at the triple bottom line in business relating to 'people, profit, planet'. To begin to answer this question, in the first instance, it is essential to define what sustainability is or is not in this clothing context. For the purpose of this discussion, sustainability will be explained in terms of its current principles relating to the social, economic and environmental consequences of our behaviour as consumers. This essay seeks to explain the significance of the fashion and textiles sector and its importance in relation to our cultural and emotional connection to clothes. This will include historical and contemporary consumption patterns (people), assessing the importance of the global industry driving macro and micro economies (profit), and outlining the physical impact this industry has, and is having, on the environment (planet). In this chapter Joan Farrer, whose doctoral thesis on these issues (Royal College of Art, London, 2000) is celebrating its tenth anniversary, sets out to clarify the core ideals of sustainability, in particular in relation to fashion and textiles, in light of the modern zeitgeist. The objective will be to review some key solutions that may offer a remedy to the current situation in order to move towards the apparent contradiction of a more sustainable fashion industry. Examples include recent research that profiles upcycling and re-manufacture (Fraser, 2009), design for source local/ sell local (Finn, 2008) and her own work in smart technological solutions for producers, retailers and consumers (Farrer and Parr, 2008). A model of 'remediation' is explored as a potential way to provide the most up-to-date solution to what remains a critical issue for the fashion and textiles industry.

WHAT IS SUSTAINABILITY?

The word sustainability has a plethora of meanings and is frequently misunderstood; unfortunately it has become synonymous and interchangeable with recycling and the environment, whereas the original rationale from the 1950s was to focus on social change to alleviate global poverty. The misrepresentation and cherry picking of values from the sustainable agenda, particularly over the last decade, by business, marketers, politicians and even by education, has led to the movement becoming hijacked for commercial purposes.

In many expert circles there is a struggle to find another word to replace sustainability, because its deeper meanings and associated philosophies have become worthless, vacuous brand development and 'green-wash' tools. One of the most cohesive descriptions given more than 20 years ago by Bruntland was that '...sustainable development is development that meets the needs of the present without compromising the ability of future generations to meet their own needs' (Brundtland, 1987). For poor countries, this was perfectly commendable and appropriate. However, as Wood points out in Chapter 5 of *Designers, Visionaries and Other Stories*:

> Before we knew where we were, we had stretched the original idea of 'sustainable development' and were talking about 'sustainable products', 'sustainable approaches' and 'sustainable housing'... politically the idea of sustainable development created the idea that there was a common agenda or consensus (Wood, 2007).

In fact at the latest count and rising there were 70 different definitions of sustainability (Holmberg and Sandbrook, 1992; Pearce et al, 1989). What do 70 plus definitions of the meaning of sustainability mean for practitioners in the fashion industry now? Which one of these 70 definitions and counting, affect the way we think about how we make and use clothes? One of the easiest visual descriptions of the complexity of sustainability is the milking stool model with its three legs and seat. Leg one represents people, leg two profit and leg three planet, which all support the seat which represents the sustainable platform. In a move towards a more sustainable fashion industry all three legs must be as good and solid as they can possibly be. Consider this example, the industry may produce an organic cotton shirt (planet) which can still be made by a child labourer (people) and flown around the globe to European markets (profit). Can we say this is a sustainable fashion product, although depending on one of your 70 definitions some would?

In North America, Europe and the UK, retailers must be 'seen to be green' due to shareholder pressure, and to be 'doing the right thing' particularly in respect of customers' brand perceptions of the ethical and environmental corporate values. However, the British fashion retail system is unique in the world, in its practice of being driven by a few huge industrial fashion retailers using a 'mono-logical' capitalist system, an overly complex and costly system of products and services, designed to relieve us of the tasks and boring repetitiveness of everyday life (Manzini, 2005).

The mono-logical system means that the flexibility required to enable a more sustainable outcome by taking business risks is difficult. British retailers also have economies of scale and can buy large volumes of clothing at ever lower prices, creating a 'churn' of affordable, well designed goods into and out of store, which can be constantly refreshed, so delighting the consumer who is ever willing to buy more. Cheaper goods mean more consumption, which in turn means cheaper goods, which means more consumption. The customer footfall into the UK industrial fashion retail stores is massive, as is the rapid turnover of stock and item sales. How can this giant industrial fashion system ever be sustainable? Also a lucrative by-product of the British system, now being replicated elsewhere, is the increasing volumes of quality clothing waste leaving the UK, traded as recycled textiles for overseas destinations. This is unregulated 'green waste', which is dumped upon other countries as any waste would be, and where the disposal of the waste at the source of consumption is not the responsibility of that country. Dumping fashion clothing on more than 65 overseas countries is eroding their local fashion textiles industries and creating profit for those few agencies handling the waste, whether private commerce or registered charities.

The major source of the problem in achieving a sustainable fashion industry is the consumer. Fashion customers are hungry for goods, yet need to feel absolved from the responsibility of the constant refreshing of their wardrobe; this is also a physical problem rather than philosophical one. Low prices, good design, good quality fashion clothing items, coupled with an exciting shopping leisure experience on the cheap, mean an increase in purchases, which is difficult to reconcile with the idea of a looming environmental armageddon. Customers can hold a large quantity of clothes in their homes stored in wardrobes, drawers, trunks, garages and lofts, but then when this storage is exhausted what can be done with these items which are almost as good as new? The consumer's only option, apart from trading on online auctions or selling to second-hand stores, is to give these fashion clothes as donations to charities, either over the counter to high street shops or clothing banks, or to the increasing number of curb-side collectors. The clothes are traded from the back of the charity shops and warehouses and sold by weight to fashion textile customers primarily in the second-hand clothing industry, to be dispersed across the globe.

Images of disappearing tribes on TV documentaries wearing branded T-shirts are a direct result of this unregulated free trade, or dumping. Second-hand clothing from the UK is recognized globally as prime quality and in particular labels, such as Marks and Spencer, are recognized as high value to the recycler, where haggling over price rarely takes place at the destination country, due to the quality of goods in each consignment from the UK (Abimbola, 2010). The more clothing generated and sold by industrial retailers, the more clothing waste is generated by the consumer and the more the UK charities benefit by donations of these clothes, which have been given both altruistically and in desperation, in order to salve the conscience of the fashion consumer in the first place. This is a vicious circle because those purged and half empty wardrobes can now contain more new clothes! In response, UK volume high street brands such as Top Shop (Treehugger, 2009), sell expensive second-hand, recycled and restyled lines alongside their main ranges, and TK Maxx (Hussey et al, 2009) offers a clothing take-back system in store, to recycle customers' donated fashion items as they purchase new ones. It is appropriate that the UK is leading the research into sustainable fashion textiles with support from government funds for data capture on the industry; examples include organizations such as the University of Cambridge Centre for Manufacturing who published *Well Dressed* (2006), because the UK is the source of this questionable business practice that other countries are following. UK educational organizations too are supporting sustainable design networks such as Textile Environment Design (TED), the Textiles Futures Research Group (TFRG) and the Slow Textiles group, and degree courses such as the BA Eco Design at Goldsmiths, University London and the MA Sustainable Fashion Design at the Centre for Sustainable Design. London College of Fashion, University of the Arts London (UAL) and the University College London Department of Anthropology are all producing items or publishing on the reasons behind and solutions to the problem. The source of the escalating crisis of what to do with and where to dump good quality fashion clothing waste is the huge industrial fashion retail and charity system in the UK. Significant research to resolve the problem on many levels should be funded by the fashion retailers and the charities together as they benefit from the lucrative world of free trade, where the consumer is a pawn.

It is crucial for the rest of the world to understand that these UK centric problems are not mirrored to the same extent overseas. We must be careful not to look to the UK for predictive and diagnostic data analysis and solutions to apply to other national systems. Each country and geographical region has its own very effective methods of dealing with the way it makes and uses clothes, which the UK has moved away from. These local methods of production and disposal should be investigated and analysed from a national perspective rather than borrowing rationale and solutions from the catastrophe in the UK where methods and data may be wholly

21

inappropriate. This is a case where the UK's northern hemisphere model of 'one size fits all' definitely does not work. Competitors in European countries such as The Netherlands and Italy, or the Antipodes for instance, operate a smaller scale boutique system in tandem with the global brands. Small retailers have the flexibility to try innovation, perhaps to make locally, using ethical trade, connecting maker and consumer, or can trade online with a first sample range, then produce the numbers in the correct sizes, almost a buy before you make and produce; a customized system. Also in the case of Australia and New Zealand, where second-hand clothing has value and material worth that is not throwaway, fashion clothing may have been passed round many times within the family and circles of friends, before being donated to charity. In the case of New Zealand, if donations remain unsold (which is often the case due to the used state of the clothing), then they will be sold to industrial scale second-hand clothing outlets such as SaveMart (Roberts, 2005) to be retailed before being dumped on Papua New Guinea (Fraser, 2009).

Today our libraries have volumes of literature describing the concepts of sustainability, rafts of business case studies, reports and illustrations, charts, mind maps, graphs and data, which help us to grapple with the complexity of the movement. But who reads these scholarly tomes? Not many fashion designers, which is a problem when it is they who are often charged with being at the heart of the consumer desire for fashion clothing, leading to overproduction of items that date, and are inadvertently responsible for the creation of fashion clothing waste. Fashioning an ethical industry has been a key driver in the UK with regard to educating the staff and student body on some associated ethical and empathetic issues, which has led to an interest in alternative production and consumption of clothing. The assumption is that this will lead to a more sustainable fashion industry and consumer.

SUSTAINABILITY IN FASHION TEXTILES

22

Fashion and textiles are integral to our culture and to our economies, indeed Schneider and Weiner in the book *Cloth and the Human Experience* assert that, 'Throughout history, cloth has furthered the organization of social and political life' (Weiner and Schneider, 1989).

Many phrases in the English language have come from this ancient industry, where cloth remnants have been found from 36,000 BP (before present) and fashion and textiles meanings are subliminally embedded into our culture. Phrases like 'after a fashion' meaning to follow a style or behaviour, 'fabric of society', 'folded into', 'text' from textile, 'tailor made', customized, a 'thread' of conversation, the list of 'material' is substantial. Historical records show the economic significance of the clothing and textiles trade. Even the trade in second-hand clothing is centuries old and in Italy in 1400, Tuscan Pawn Brokers kept detailed accounts showing that 40 per cent of trade was in clothing and apparel. So too, documents from the late 1500s in the UK prove second-hand textiles and clothing to be a significant feature of economic development. During the 1500s and 1600s the banking sector was in its infancy in Europe and small coins were a rarity. As garments and textiles were an investment and the items often carried great value, they were used as barter, where they were seen as part of a trading system to replace cash. This was a textile society where fabric was an alternative currency, particularly for the working poor, although both the wealthy and poor in society used textiles and clothing to replace hard cash.

The late 1500s saw the development of 'small thrift' economics, cottage industries and the emergence of the solo female trader who began to benefit from the transformation of a society of relative scarcity giving way to one of plenty in commodity terms. The 1600s and 1700s saw a wide array of general goods for sale and barter with an increase in the volumes of clothing and textiles because of the importation of cheaper cotton from India, leading to a greater volume of cheaper clothing and household textiles to trade and an increase in dealers. The growing abundance of goods enjoyed by the working classes ensured increased material flows that meant that the trading of old goods, clothing and textiles for new items gathered momentum, expanding commercial activity and developing significant micro-enterprises. Textiles knowledge at the time was striking, as was the scale of trade in these businesses. This is explained by Lemire:

> ...the technologies of textile use and reuse were foundational aspects of European society and economy for generations. Skills in reuse were practiced in many areas of society from households through to retail marketplace, with intersecting networks of use and exchange. The trade offered marginal commercial opportunities for poor women and minorities, while being a staple of various retail and merchant sectors (regional, national and international) (Lemire, 2010).

During the 1800s, with the development of industrially produced textiles made of cotton, wool and flax, the price, and standing, of textiles in society began to decline, a trend that has continued to the present day, where fashion clothing in the UK is a disposable commodity. Then, common people were swapping clothing for goods such as china, which resulted in the increase of quantities of used apparel being shipped overseas to places such as Poland and The Netherlands. However, the upper classes had been moving away from this trade in second-hand textiles, which had a decreasing value, and were pawning jewellery and fine goods to raise money instead. Consequently, the second-hand textiles and clothing trade became synonymous with the working classes, making decency, style and comfort possible to a certain extent for the poor, where quality goods, in particular tailored clothing, were sought after in the second-hand market, thus creating a distinction between the classes of those who wore new and those who wore second-hand. Fine sewing was the domain of the tailor and seamstresses, who repaired and altered items (including those involved in refurbishing hosiery) in order to adapt clothing successfully for those who could pay. Such transferable skills with regard to making, altering and repairing clothing were passed down through generations, although this began to change with the introduction of mechanization, development of industrially manufactured clothing, the introduction of the domestic sewing machine and the emergence of the mail order, ready to wear industry, during the early 19th century.

Nevertheless, the second-hand clothing market continued to flourish and was supported by the culture of home sewing in the UK, because of the cost of new items. There were periods of renewed interest in the field of remake in the 1900s particularly during and after World War I and World War II when new clothing and materials were scarce. Towards the end of the century, from the 1990s, there has been a gathering momentum for remake and upcycle, with individual designers in that decade leading the way under the genre of deconstruction. One such master of changing the way we think and use clothes is Margiela, whose contemporary use of raw materials is sourced from second-hand or army surplus commodity clothing, which has some of the lowest exchange values in the second-hand clothing system. However, Margiela moved the notion of second-hand away from its associations with the poor and working classes, repositioning the garments at the top of a hierarchy of prestige. He converted textiles and clothing waste into something of desire with a high commercial value and a new understanding of material worth. He is the antithesis of the 19th-century second-hand clothing workers or rag pickers discussed by Quennell in 1964, who were 'deemed to be the lowest and weakest of citizen who were scavengers, rag pickers and peddlars [sic]' (Quennell, 1964).

In her essay *'The Golden Dustman: A critical evaluation of the work of Martin Margiela and a review of Martin Margiela Exhibition (9/4/1615)'*, Evans also draws together numerous historical references affirming the low status of the second-hand clothing market, its workers and consumers in the past and she argues, in the light of Margiela's artistic genius, for a change of status for the industry, referencing previous scholarly texts written on the subject:

23

Benjamin goes on to comment on the analogy that Baudelaire made between the rag picker and the poet – for which latter term we could as well substitute 'artist'. Or, as I [Evans] would argue in this context, 'fashion designer'. Not, of course that Margiela's status as a fashion designer is low: on the contrary. But his interest in scavenging and revitalizing moribund material is not that dissimilar to that of Baudelaire's poet/rag picker (Evans 1998).

The 1990s saw the remake, deconstruction and upcycling of clothing, either in its most beautiful form from avant-garde designers such as Galliano or Chalayan or a mid-range quality such as Ann-Sophie Bach and Jessica Ogden to the often rough but acceptable work of students and 'cool' home sewers.

A GLOBAL INDUSTRY DRIVING MACRO AND MICRO ECONOMIES (PROFIT)

The world textile and clothing trade reached US$530 billion in 2006 (WTO, 2006). Today fashion and textiles continue to drive economies throughout the world on micro and macro scales, from a multitude of developing world craft producers to industrial fashion manufacturers and retailers. In Europe, the textiles and clothing industry turned over €211 billion in 2007, produced by approximately 145,000 enterprises employing more than 2.5 million people; the small and medium enterprises (SMEs) employed an average of 1–9 people each (European Apparel and Textile Confederation, 2008). The value of the global industry is estimated to reach US$1781.7 billion by the end of 2010 and in excess of 30 million people work in the industry in China alone. The European Union continues to dominate global markets and represents the world's second largest exporter of textile products after China (ObservatoryNANO, 2009). The Industry has focused on 'start of pipe' streamlining, design and branding and optimizing supply chain management, aiming to supply consumers with high added value low price products at very short notice. Based on current production, consumption and disposal business practices the diagram shown in Figure 1.1.1 illustrates the complexity of existing fashion supply chains. The difficulty that the fashion industry faces, in order to supply a future sustainable and ethical customer, is how to alter its philosophy and multiple business models whilst remaining profitable. This diagram shows the typical process for development and manufacture of a fashion textile product, commencing with fibre processing, through textile manufacture, garment assembly, distribution, sales and eventual disposal. Most processes could be local but are usually global. This flow chart also points out the various chemical inputs required throughout the manufacturing process that are usually not associated with the finished product. Through introducing 'use' and 'disposal' as the follow up phase to 'distribution' and 'sales', detergents can be viewed as chemical input, further adding to the complexity of the issues faced by the fashion industry (Farrer and Fraser, 2010).

THE PHYSICAL ENVIRONMENTAL IMPACT OF THE FASHION TEXTILES INDUSTRY (PLANET)

As an industry, textiles and clothing was the core driver of the Industrial Revolution in Great Britain, the developmental effects of which cascaded through Europe, the Commonwealth countries and North America over a 200-year period from the late 1700s. Now the industry is again at the forefront of a revolution, in technical terms, with the development of smart polymers and nanotechnology to produce functional well-being performance fashion textiles products with minimum waste, which is recyclable. But perhaps more importantly, this time

24

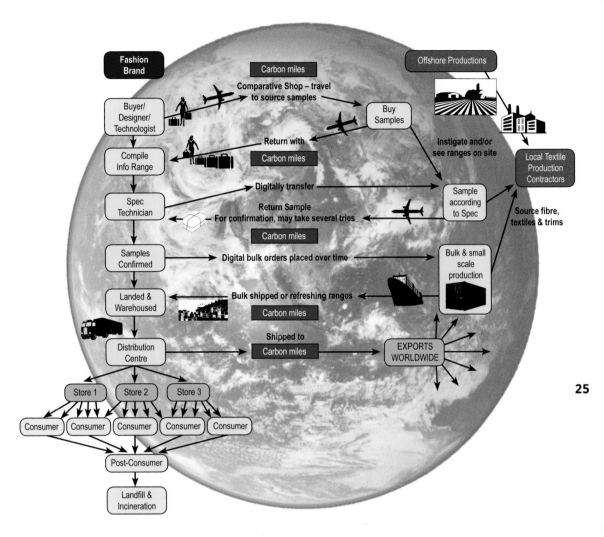

ABOVE | Figure 1.1.1 Fashion and textile typical supply chain designed by Farrer and Fraser (2009)

round, fashion, like the organic food industry, is having a humanizing effect communicating the complex technical issues central to a sustainable philosophy involving production, consumption and waste. Through the emotive medium of clothing, which anyone of any age and culture can relate to, fashion is visualizing and communicating positive and negative issues around the way we make and use clothes.

For the individual in Europe and the northern hemisphere, until recent times, the environmental crisis has been imperceptible. The sun rises and sets, the birds sing, children are born, and life continues as it did yesterday and yet crisis there is, inextricably linked to commerce and commodity production of fibres. This is seen many times in the case of cotton production, which is harmful to ecosystems, such as the 1990s Aral Sea desertification, water and soil contamination in Russia, and post-2000 in the Mackenzie River delta, in Queensland, Australia. Free trade and the generation of money-capitalism have become destructive to the ecosphere, which has exquisite limits, unlike the generation of capital (profit), which is limitless, and the

resulting globalization and neo liberalism of fashion brands has heightened the exploitation of nature (planet) and labour (people) in recent years. No one can deny fashion clothing's seductiveness, through the fact that its industries produce a feeling of luxury and wealth, a feel-good factor that is affordable, easily reinvigorated and deeply cool.

THE CASE STUDIES

The case studies illustrate examples of remedial action where the objective is to review key solutions that may suggest a remedy to the current unsustainable situation in the fashion clothing industry.

- Case Study 1: Source, upcycling and remanufacture for waste materials recovery.
- Case Study 2: Source local and sell local.
- Case Study 3: Communication of the sustainable agenda and its issues through fashion clothing.

Case Study 1 – Kim Fraser
'Redress – ReFashion as a Solution for Clothing (Un)Sustainability'

Murray in the book *Zero Waste* explains what upcycling textiles from fashion textiles waste should be about:

> not merely conserving the resources that went into the production of particular materials, but adding to the value embodied in them by the application of knowledge in the course of their recirculation (Murray, 2002).

The fashion industry epitomizes unsustainability with its fast changing trends, high minimums and planned obsolescence, contributing millions of tonnes of clothing to landfill, incineration and third world dumping. One solution is to ReFashion – a process that intercepts discarded clothing (post-consumer textile waste), reclaims, re-cuts and refashions, returning the item to the clothing stream, effectively creating a new loop, postponing its grave ending, thus reducing both textile waste and the demand on raw materials required in the manufacture of new textiles.

The research of designer and maker Kim Fraser was to promote debate and alter perceptions of second-hand materials and ReFashion concepts. Her thesis took the standpoint that discarded clothing is an untapped commodity, a rich fibre/textile resource to be conserved and transformed into contemporary fashion. However, while the word ReFashion has filtered into fashion product terminology and many fashion labels are reusing and recycling second-hand and vintage fabrics, it remains a 'one-off' principle. The 'multiple production of one-offs' that arises from the haphazard and diverse nature of the starting point materials (Dunn, 2008), is an approach to manufacture that is suitable for low unit quantities and niche markets, but is unlikely to be attractive to large fashion businesses. Through reflection on the deconstruction/reconstruction process of developing prototypes, issues involved in the current practices of 'materials recovery' in the secondary textile industry were identified. As the work developed the investigation sought to determine whether the current ReFashion process could be suitably scaled up for larger fashion businesses. Typically, large fashion businesses operate in an 'economy of scale', that is, large companies lower the average cost per unit through increased production. The subsequent advantage is that buying bulk is cheaper on a per-unit basis, allowing a producer's average cost per unit to fall as scale is increased. The implication, if ReFashion were to begin to address effective manufacture and parallel this bulk model, was an obvious need to source a bulk supply of input stock.

Fraser's studio investigation targeted specific post-consumer textile waste and established several processes unique to ReFashion. During the first scoping phase of the ReFashion experimentation,

27

LEFT | Figure 1.1.2 ReDress – 'T series' dresses on rack designed by Kim Fraser. Digital photograph, Band (2010)

numerous discarded items were transformed. One of the early experiments, which transformed a men's trouser into a contemporary fashion dress, highlighted the potential for targeting men's dress trousers: quality of cloth within a semi-standard size and shape; similar construction and tailored details; apparent availability of large quantities. The strategy to limit input stock to men's trousers permitted a way of filtering the complex issues created by worn garments, providing opportunities for in-depth examination and interpretation. Throughout the research, focus was placed on process rather than artefact, processes used in the development of prototypes were analysed and tacit understandings scrutinized. The emphasis was on repeatability, identifying and recording the appropriate manufacturing knowledge, techniques and processes that would be required to repeat the 'product'. In this manner issues relating to patterning for repeatability were highlighted (disassembly, usable piece size, nature of the second-hand garment, complexity and jigsaw fit, expertise of the cutter, initial selection and stock recording).

Fraser's practice-based research highlighted the new skill set required, revealing potentially achievable methods to manage the processes that were determined unique to ReFashion. Through comparative analysis, the identified processes were measured against standard garment assembly procedures within current manufacturing systems to determine the required modifications. Potential adaptations of a manufacture process for ReFashion have been identified and documented, with the T-series (Figure 1.1.2) providing evidence of the possibility of ReFashioning a standardized fashion 'product'.

Case Study 2 – Angie Finn
Full Circle: A Collection of Prototypes

28

The research of designer and maker Angie Finn engages with another one of the key issues facing the fashion textiles industry in terms of future sustainability: that of the well-being of fashion industry workers in Australia and New Zealand (people). Her honours dissertation (completed in New Zealand in 2008), and subsequent publication and conference presentations, examine the contribution that design can make to sustainable manufacturing; particularly design for local production and consumption. An important aspect in this discussion of source, the work suggests that the 'made in China syndrome' (as she refers to the current state of over-consumerism in Australia and New Zealand) could be bought to a close through design to minimize waste and maximize opportunity for 'people': in this case both garment workers and the SMEs that employ them. Her work is interesting in that it focuses on a local approach that could be put into practice by a framework of SMEs that already exist. In addition the design process is highly transferable and could be put into practice almost anywhere with minimal set up costs and a design ethos that progresses at the same pace as the skills of workers. This collection is a physical and conceptual embodiment of a source local/sell local approach.

In a different approach from the previous case study, where reuse/recycle/reclaim methods were applied to the problems of fashion textiles waste and explored through ReFashion processes, Finn's ideas engage with stopping these problems at the source. She argues, with Farrer (Farrer and Finn, 2009, 2010) that this is not an unrealistic ideal and is in fact possible through the development of a sustainable industry, in the truest sense of people, profit and planet, through adoption of a design process model that stops the waste at source by making better use of the raw materials and labour involved in making fashion garments. Finn identifies that much discussion surrounding fashion textiles sustainability is focused on environmental sustainability, raising the issues and problems but failing to offer what she refers to as real world solutions for small business.

Finn examines one possible solution in her honors dissertation entitled *'Fashion manufacturing in New Zealand: Can design contribute to a more sustainable industry?'* (Finn, 2008). Her final

collection, entitled Full Circle: A collection of prototypes, showcases a range of highly desirable and commercial products that could be manufactured by SMEs in New Zealand (or indeed in Australia or the UK) in smaller quantities for a realistic market price. Figure 1.1.3 shows one sample from the collection. The inspiration behind the collection was to design a series of prototypes that minimized fabric usage, could be constructed using a straight sewing machine, cut without a pattern and fit without the requirement of a pattern grading process. The aim of this research was to demonstrate that design could be used as a practical process to recreate jobs and re-energize the career of making. As she explains:

> It has been, and continues to be, of real concern to me that perhaps Australia sold out far more than we realized for the price of cheap T-Shirts and jeans. I worked in the industry in Australia in the late 1980s, before the import tariffs were lifted, with a couple of hundred factory workers who were getting on with their lives, paying their mortgages, catching up for a few drinks on Friday after work, bringing up young families. I was very young myself but remember thinking...this [lifting the tariffs] is a really stupid idea, and so it was. It seemed that the industry died overnight, trying to compete on price. The great thing about the fashion industry is that it can change, reshape itself and re-grow. This is our strength; we can work with what we have rather than what we have not (Finn, 2007).

A key factor in Finn's interest in sustainability is to improve the well-being of individuals, particularly through a reconnection between the user of a product and the maker. Although the focus of this research appears to centre on people and profit, as the next case study will show, this kind of source local/sell local approach also has great benefits in terms of environmental sustainability.

LEFT | Figure 1.1.3 Finn, A. (2008). Hooded jacket and dress designed by Angie Finn. Collection: Full Circle

29

CONsCIENCE CLOTHING shoot GARMENT MILES

SABRINA	SANJIV	SAM	JACKSON	SALLY
22, 149	+ 43, 765	+ 22, 426	+ 15, 256	+ 3, 296

Total CO2 emission approx = 50.98 TONNES

ABOVE | Figure 1.1.4 Conscience Clothing shoot, showing garment miles to New Zealand and related carbon footprint. Life-size posters designed by Joan Farrer. Digital photographs: Parr (2008)

Case Study 3 – Joan Farrer
Conscience Clothing: Communicating Through Fashion Clothing to Inform the Sustainable Consumer

The research of Joan Farrer is to understand the complexities of the sustainable agenda in terms of industrial fashion and textiles design, manufacture and disposal and then to communicate these issues to all consumers within this chain leading towards knowledge building and behavioural change. Manzini (2005) talks about enabling and disabling solutions (related to design, production and sustainability) and his argument is that human ability to gather, learn from, and utilize knowledge to apply to situations, including design and production and consumption, is now diminished because of advances in mechanization. This has become a disabling phenomenon, accelerating the loss of essential skills and knowledge transfer such as how to design, how to consume, how to make and reuse clothes, how to dress, what to wear and most importantly what to accept from a consumer point of view (Farrer and Fraser, 2009). Sourcing and manufacturing garments offshore for cheap sales in the home market, appears to be a sensible and inexpensive process, but if the social, economic and environmental cost is not effectively incorporated into the retail price of the garment or its effects not communicated properly to the consumer, we are supporting the concept of 'out of sight; out of mind' in terms of well-being for workers, environmental pollution and natural resource exploitation. This suggests a greater need than ever before for transparency and communication in the supply and disposal chain of fashion garments, to inform consumers and to enable the right purchasing choices to be made. Fashion consumer awareness campaigns such as the second collection of 'Conscience Clothing' (Farrer and Parr, 2008), targets a new breed of sustainable consumer. In this work, five students were chosen at random, and photographed in their everyday clothes. A note was made of the point of origin or 'made in' label, sewn into their clothing and printed on their footwear. A calculation was made of each garment's mileage from the garment manufacturing labelled point of origin to New Zealand, and the mileage was included on the poster. Using the Qantas air miles carbon converter, estimates were made of the total garment miles, an astonishing 1,069,112, with carbon emissions of 50.98 tonnes. A series of life size images were digitally printed on fabric banners and exhibited in Auckland. From the five student models, the poster 'Sally' illustrates local clothing sourced from Australia and New Zealand which is calculated as travelling 3296 miles, compared with 'Sanjiv', who illustrates the consequences of clothing sourced from parts of Asia with a total of 43,765 miles. This highlights the potential benefits, in carbon terms alone, of close to market design and manufacturing, but poetically illustrates and communicates complex information, in a simple way, for consumers to make educated choices to support behavioural change.

CONCLUSION

The multiple sources of fashion textiles production coupled with the enormous number of definitions of sustainability mean that there are too many variables in creating a sustainable fashion and textiles sector in mass market terms. It is a Utopian ideal. For instance, a limitation to the wider adoption of sustainable philosophies in the sector is related to the cost of more sustainably produced products from accredited supply chains with an integrated product policy. There are concerns from industrial manufacturers that there may be low compatibility of new 'sustainable' production processes with current production processes and, critically, insufficient production capacity of successful new production methods, as experienced in the organic food industry, where supply and demand have been an issue along with fraudulent claims of product authenticity.

31

More sustainably produced commercial fashion clothes that factor in the 'true cost' of these products from an environmental and social perspective are a higher cost than traditionally manufactured and priced lines, not only in the area of equipment, but in hiring expertise in sustainable practices, use of more expensive, less damaging raw materials and considered disposal or upcycling of fashion clothing waste. The industry is dominated by SMEs, the fragmentation of which complicates the implementation of new developments, theories and processes. Most companies are not informed about what is being done or discussed in literature on sustainability or aware of research centre R&D. In addition they have limited resources to invest in new technical developments and business processes unless forced to by legislation. The fashion textiles sector is mainly cost driven; this move towards a more sustainably sourced product will be realized in the final product cost to the consumer, something that is not feasible at the moment due to cost.

However, a polarization is taking place in the fashion textiles industry (Farrer and Fraser, 2009) which means that there is an opportunity for an alternative sector to emerge from co-creators of generation C (Pearce and Young, 2007), where a political message is being delivered to makers and users developing new kinds of business models for distribution and sales. As outlined and discussed here, methods of production and consumption from the raft of sustainable definitions, such as make local/sell local (Finn, 2008) and deconstruct/reconstruct (Fraser, 2009) are slowly being understood. Small-scale business in this arena is producing fashion clothing design which is artisanal and underpinned with craft skills, where the handmade and its 'one-off' idiosyncrasies are celebrated. These garments are becoming 'sustainable fashion badges' worn by an educated consumer who wears the small-run commercial or handmade garment as a political flag in order to rail against the unsustainable industrial fashion clothing system in the developed world.

REFERENCES

Abimbola, O. (2010) *Igbo Trade Networks & Secondhand Clothing*. Recycling Textile Technologies Workshop, University College London Department of Anthropology

Benjamin, W. (1997) *Charles Baudelaire: A Lyric Poet in the Era of High Capitalism*, Verso, London and New York

Brundtland, G. H. (1987) *Our Common Future*. World Commission on Environment and Development. www.worldinbalance.net/pdf/1987-brundtland.pdf (accessed 31 August 2008)

Dunn, J. (2008) *ReFashion reDunn*. Masters Thesis. Massey University,Wellington. http://muir.massey.ac.nz/handle/10179/711 (accessed 16 June 2009)

European Apparel and Textile Confederation (2008) *The EU-27 Textile and Clothing Industry in the year 2007*. Euratex www.euratex.org/system/files/.../Economic+Situation+2007-Marchi.pdf (accessed 27th August 2010)

Evans, C. (1998) 'The Golden Dustman: A critical evaluation of the work of Martin Margiela and a review of Martin Margiela Exhibition (9/4/1615)', *Fashion Theory*, vol 2, no 1, pp73–94

Farrer, J. and Finn, A. (2009) 'Full circle: The future of sustainable fashion manufacturing in New Zealand', in *International Foundation of Fashion Technology Institutes (IFFTI) 2009: Fashion and Wellbeing*. The Centre for Learning and Teaching in Art and Design (cltad), London

Farrer, J. and Finn, A. (2010) 'The power of a single prototype: Sustainable fashion textile design and the prevention of carcinogenic melanoma, in P. and T. H. Bartolo (eds) *International Conference on Advanced Research and Rapid Prototyping: Innovative Developments in Design and Manufacturing Advanced Research in Virtual and Rapid Prototyping*. Boca Raton and London

Farrer, J. and Fraser, K. (2009) 'Conscience clothing: Polarisation of the fashion textile market', *Textiles – Quarterly magazine of the Textile Institute*, vol 2009, no 1, pp10–13

Farrer, J. and Fraser, K. (2010) 'Sustainable v unsustainable: Articulating division in the fashion textiles industry', *Antipodes Design Journal,* vol 1, no 1, pp1–18

Finn, A. (2007) 'Ethical fashion: The human stories driving a fashion movement', *Fashion in Fiction: A Transdisciplinary Conference*. University of Technology Sydney

Finn, A. (2008) 'Fashion manufacturing in New Zealand: Can design contribute to a sustainable fashion Industry?' http://eprints.qut.edu.au/31512/ (Accessed 14 June 2010)

Fraser, K. (2009) 'ReDress: Refashion as a solution for clothing (un) sustainability', unpublished master's thesis. AUT University, Auckland, New Zealand

Holmberg and Sandbrook (1992) cited in C. L. Davey, A. Wootton and C. T. Boyko, (2005) *Synergy City: Planning for a High Density, Super-symbiotic Society*. http://linkinghub.elsevier.com/retrieve/pii/S0169204607001466 (accessed 7 July 2010)

Hussey, C., Sinha, P. and Kelday, F. (2009) 'TK Maxx. Textile and clothing recycling and reuse is an under researched field', in *Responsible Design: Re-using/Recycling Clothes. European Academy of Design International Conference*. www.ead09.org.uk/Papers/032.pdf (accessed 7 July 2010)

Lemire, B. (2010) *The Second-hand Textile Trade in Europe & the Atlantic World: Practice and Enterprise, c. 1600–1850*. Recycling Textile Technologies Workshop, University College London Department of Anthropology

Manzini, E. (2005) 'Enabling solutions, social innovation and design for sustainability'. http://195.157.47.225/mt/red/archives/2005/09/ (accessed 17 November 2006)

Murray, R. (2002) *Zero Waste – zero waste cover & contents*. Greenpeace Environmental Trust Publishers, www.ecologycenter.org/iptf/recycling/zero%20waste%20murray

ObservatoryNANO (2009) *Technology Sector 10 Textiles report*. www.observatorynano.eu/project/catalogue/2TE/ (accessed 27th August 2010)

Parr, H. (2008) Conscious Clothing shoot, showing garment miles to New Zealand and related carbon footprint.

Pearce, J. and Young, S. (2007) 'Ch-ch-changes', *Idealog*, 14, 69. http://idealog.co.nz/magazine/march-april-2008/

features/ch-ch-changes (accessed 18 June 2008)

Pearce et al (1989) cited in C. L. Davey, A. Wootton and C. T. Boyko, (2005) *Synergy City: Planning for a High Density, Super-symbiotic Society.* http://linkinghub.elsevier.com/retrieve/pii/S0169204607001466 (accessed 7 July 2010)

Quennell, P. (ed.) (1964) *Mayhew's London: Selections from London Labour and the London Poor.* Spring Books, London

Roberts, S. (2005) 'Auckland recycling industry study, a survey of recycling and second-hand businesses in the Auckland region'. Envision New Zealand. www.wasteminz.org.nz/conference/conferencepapers2005/Sarah%20 Roberts.pdf (accessed 8 September 2008)

Textiles Futures Research Group (TFRG) www.tfrg.org.uk/magazine/current (accessed 7 July 2010)

Textile Environment Design (TED) www.chelsea.arts.ac.uk/22072.htmn (accessed 7 July 2010)

Treehugger (2009) 'Top Shop... new label launching at Topshop's London flagship Oxford Street locations, My Only One Remakes Vintage Clothing into Retro Street Wear'. http://treehugger.com/files/2009/08/my-only-one.php (accessed 7 July 2010)

University of Cambridge (2006) *Well Dressed? The Present and Future Sustainability of Clothing and Textiles in the United Kingdom.* Cambridge. www.ifm.eng.cam.ac.uk/sustainability/ (accessed 25 June 2009)

Weiner, A. B. and Schneider, J. (1989) *Cloth and Human Experience.* Smithsonian Institution Press, Washington and London

Wood, J. (2007) 'Relative abundance: Fuller's discovery that the glass is always half full', in J. Chapman and N. C. Gant (eds), *Designers, Visionaries and Other Stories: A Collection of Sustainable Design Essays.* Earthscan, London, ch. 5, pp96–119

WTO (World Trade Organization) (2006) *International Trade Statistics 2006,* available at www.wto.org/english/ res_e/its2006_e/itsOb_toc_e.htm (accessed December 2006)

1.2 CASE STUDY
UPCYCLING MATERIALS FOR FASHION

Materials are routinely selected for aesthetic and functional reasons but a fashion designer can choose to work with recovered materials, which positively contributes to the management of textile waste. Australian fashion label Romance Was Born regularly work with reclaimed fabrics for its fashion collections. The Sydney-based fashion label, founded by Anna Plunkett and Luke Sales, is revered for its use of unusual and eclectic materials, handcraft techniques and eccentric, playful styling. The Garden of Eden collection, worked in collaboration with Del Kathryn Barton, included craft techniques and patchwork fabrics, over-the-top embellishment and found objects. This signature style enables the designers to provide their customer with a uniquely individual garment, a one-of-a-kind piece that is created as an item to be treasured.

Broken, damaged and redundant objects can be refashioned and value added through the process of upcycling, which in the work from Plunkett and Sales relies on the use of decorative textile techniques such as printing and embroidery. As a sustainable strategy for design, upcycling provides a designer with the opportunity to reassess the real worth and value of a waste material through the design and manufacture of new products. Rather than recycling, which can result in the downgrading of a material, informed designers are engaging in strategies such as upcycling to further prolong the life and value of a product and material. This reuse and repurpose of existing materials allows a designer to divert textile matter away from incineration or landfill.

Much of the reclaimed or recycled fabrics used will come from pre-consumer and post-consumer waste and Jana Hawley provides a broad account of the textile recycling system later in this book in Chapter 4. Pre-consumer waste is the refuse material generated during the manufacture of textile products, while post-consumer waste is thought of as pre-worn, manufactured garments that are sourced through second-hand clothing merchants and charities. By making use of these resources, a designer can remanufacture fragments or lengths of cloth to create original, one-off garments. But this challenges the notion of a standardized fashion garment or collection. When working with reclaimed materials it becomes impossible to standardize one garment into a set or series since material supplies are irregular and quantities unpredictable. Furthermore, there are technical considerations to bear in mind and overcome when working with recovered materials. For instance, a designer needs to be mindful of the condition of the raw materials noting stains, holes or areas of fraying, while also working out a method for the careful deconstruction of an existing garment. To some designers these issues may seem too complex or difficult to consider the approach worthwhile, while to designers such as Plunkett and Sales these factors can be the catalyst for new ideas. For both the designer and the wearer ultimately it is the exclusivity associated with the remanufactured fashion garment that becomes the attraction, aside from the environmental benefits, and perhaps that is why upcycled garments continue to hold their popular appeal.

LEFT | Figures 1.2.1 & 1.2.2 Romance Was Born, *Renaissance Dinosaur*, S/S 2010. Image courtesy of Little Hero, www.romancewasborn.com

OVER | Figure 1.2.3 Romance Was Born, *The Garden of Eden*, with Del Kathryn Barton 2008. Silversalt Photography

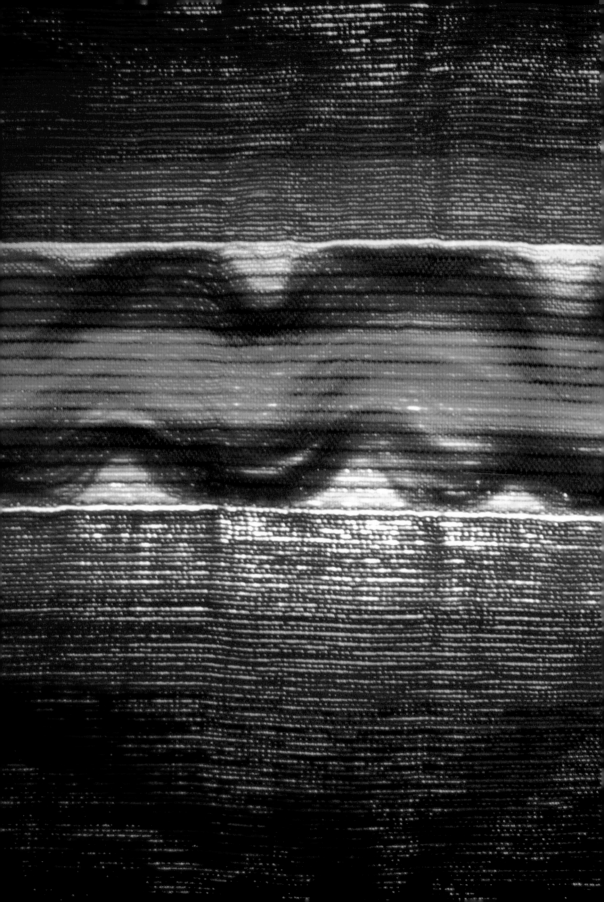

1.3 CASE STUDY
NEW MATERIALS FOR FASHION

Fashion designers have been increasingly reliant on the appropriate selection of materials as an approach to sustainable fashion. Garments can be made from renewable or biodegradable fibres, reclaimed materials or materials created through new technologies. While we know that often the most significant sustainability impacts related to clothing are created through laundering and drying, materials nevertheless play a significant role in moving towards more sustainable fashion practices. The potential for new textile materials to provide solutions for sustainable fashion has been little explored in the fashion industry. However, high performance materials can maximize garment durability, while a garment can be designed to exploit fabric ageing: these ideas and more show that through alternative material selections a garment can be designed with an extended lifecycle in mind. Furthermore, fashion can be created to adapt to different environments, climates and situations, through simple or complex transformable techniques.

Textile designer Jennifer Shellard explores the use of technology in conjunction with traditional craft skills in her experimental textile pieces. In the piece entitled Transitions II (Figure 1.3.3), Shellard directs an external computer-animated light to change and enhance a gradated coloured strip that is integrated within a hand woven material base. The gradual colour change in the strip is slow and measured and the viewing experience is both intriguing and meditative. Shellard's abstract approach to textiles demonstrates the convergence between craft and technology.

At the same time her work opens the door to alternative conversations about materials and their appropriateness to fashion. These conceptual approaches could lead designers to think about the possibility of new textile materials for garments that engage or transform. Garments that can change, adapt or evolve may encourage a relationship between wearer and garment that is much deeper than can be achieved through typical fashion solutions. And it is this connectivity to fashion that can help in the reduction of clothing consumption.

A central problem with fashion is that often a garment is disregarded before it ceases to function. In the case of a fashion garment this can relate to meaning. A garment can be disregarded because it no longer answers a perceived 'need'; and essentially the 'need' here is an emotional one. Sustaining a wearer's interest and engagement with a garment is then the real challenge. However, if a designer can create a garment that can adapt and transform, and reflect the wearer's invested care, then we can begin to rethink our engagement with our clothes.

LEFT | Figure 1.3.1 Experimenting with light responsive woven material and encapsulated within a polyamide monofilament double cloth channel, photographed under UV light. Jennifer Shellard, 2004 www.fashion.arts.ac.uk/25989.htm

OVER LEFT | Figure 1.3.2 Free standing installation consisting of panel woven with spun silk and light responsive elements, tensioned over frame with concealed UV light. Jennifer Shellard, 2005 www.fashion.arts.ac.uk/25989.htm

OVER RIGHT | Figure 1.3.3 Close-up stills from projected light transition showing gradual colour change on woven textile from warm to cool stripes. Jennifer Shellard, 2009 www.fashion.arts.ac.uk/25989.htm

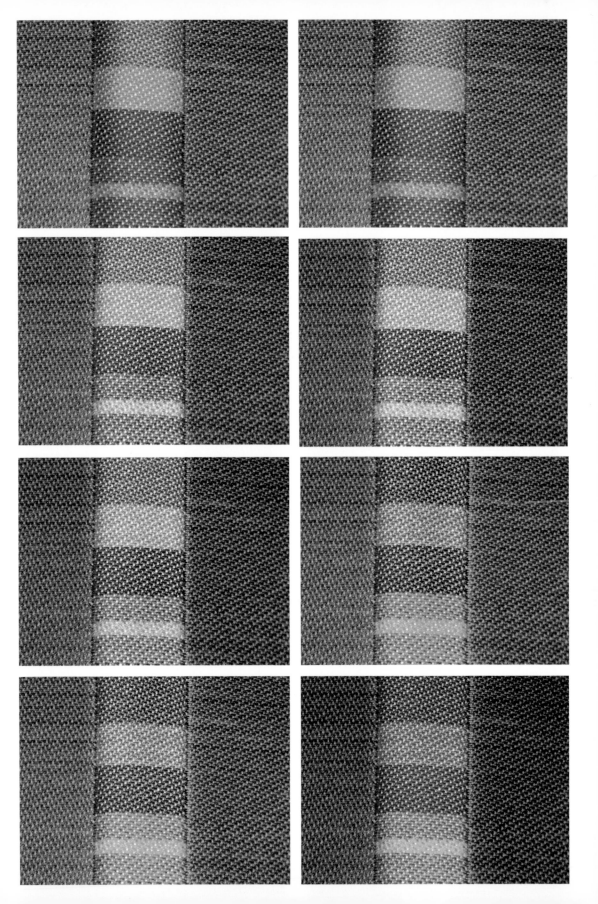

MARIE O'MAHONY

1.4 SUSTAINABLE TEXTILES:
Nature or Nurture?

'WATER... IS THE NEW OIL' (GRIMOND, 2010)

There was a time, in the not-too-distant past, when the question of textile sustainability centred on water and oil, natural and synthetic. Essentially natural fibres were generally regarded as good, coming from renewable resources and capable of being recycled. Synthetics, on the other hand, have been seen as largely derived from oil, not a quickly renewable resource and more problematic to recycle. This viewpoint is changing and one of the main reasons is undoubtedly water.

Farming is an ancient occupation and farmers have over the centuries managed their crops, often against great environmental odds. Stone carvings found inside the Egyptian pyramids depict early farming practices, showing the farmers growing a range of crops despite the unpredictability and reliance on the flooding of the Nile. So what has changed? Over the last 60 years the world population has grown from around 2.5 billion to almost 7 billion. The net result of this is that irrigated land for farming purposes has doubled while water drawn has tripled. Water usage varies sharply depending on the climate, soil and the crops being grown. British farmers use just 3 per cent of water withdrawals, while Indian farmers rely on irrigation for almost 90 per cent of their needs. Population growth and its incumbent demand for more food and clothing have certainly impacted on farming. However, much of the problem can be traced to environmental mismanagement at a national and international level. In reference to the impact of cotton on the Aral Sea, the World Bank states it bluntly: 'Under the Soviet system, economic growth was pursued in blatant disregard to natural conditions' (Grimond, 2010). In this well documented environmental disaster, the problem can be traced back to a deliberate decision to divert the rivers needed for irrigation away from the Aral Sea to the desert in order to produce cotton, rice and fruit. Today, Uzbekistan is one of the largest exporters of cotton. Quite an achievement for a country unsuited to growing the crop.

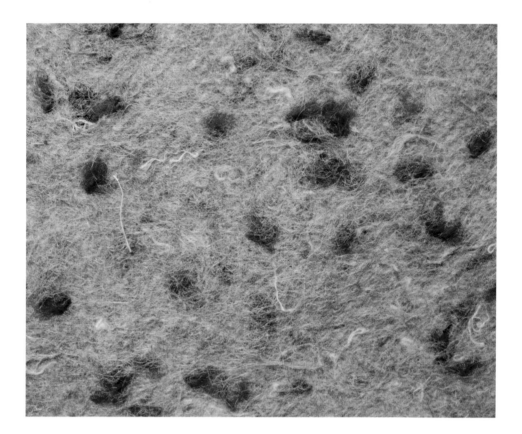

ABOVE | Figure 1.4.1 Natural fibres are being recycled for applications in a number of industries such as automotive interiors as shown in this example of recycled wool carpet and jute from LaRoche

Oil forms the basis for many of the man-made textiles currently in use. Concern about their environmental impact has led to a re-evaluation of all aspects of the industry, from yarn production, textile coatings and finishes through to product design, use, reuse and disposal. Water usage is one area where synthetics find that they can offer an advantage over natural fibres, particularly during the product's lifespan. With an increasing trend towards the use of metered water not just in industrial but also in domestic use, consumers are becoming more aware of how much they use and the thirstiest domestic appliances. Laundry consumes almost as much as showers and baths and is one area where regulation is being applied. Drought-prone Australia has initiated the Water Efficiency Labelling and Standards Scheme (WELS, 2010), requiring selected water-using products such as washing machines to have a star efficiency rating. The Government Department of the Environment, Water, Heritage and the Arts estimates that an efficient washing machine can use as little as one third the water of an inefficient model. One of the benefits of synthetic over natural fibres is lower water and energy usage during the laundry process. The frequency of laundry is also being considered, particularly with the advent of new coating and finishing treatments for synthetic materials.

While neither natural nor man-made fabrics are proving to be perfect, both have positive attributes to offer in the search for more sustainable materials. The question that material scientists and designers are now asking is whether these strengths can be developed into new, possibly hybrid materials. Can the future be a synergy of natural and man-made, nature and nurture?

NATURE

> Most architects and designers... feel that high technology is bound to disturb the ecological balance... they express this concern through a nostalgic longing for the past, in an attempt to return to a simpler, more primitive way of life. Yet one cannot turn back the clock, however good the reason may seem (Papanek, 1995).

Natural fibres are generally divided into two categories, those derived from plants such as cotton and hemp, and those derived from animals that include wool and silk. Cotton, wool, silk and linen have long dominated the market for natural fibres. More recently a new group of natural fibres has been launched on the market that includes wood and bamboo. These offer new aesthetics and performance characteristics and are being used in industries ranging from clothing to composites, a process where two or more materials are combined to produce a third, new material with enhanced performance characteristics. Many of these materials have been used for some time in traditional textiles. However, the development of advanced processing technologies is making it possible to produce them commercially. There is no longer a widespread assumption that because a material is natural it is going to be environmentally friendly. Natural fibres must now prove their environmental credentials in just the same way as their man-made counterparts.

In considering the sustainability of any fibre, production is just one part of the story. In the case of cotton it is a big part (discussed later in this chapter) but laundry care during its lifetime is also a major issue. The Cotton Incorporated Lifestyle Monitor™ has found in its survey that more than 60 per cent of women prefer to wear cotton trousers that are wrinkle resistant. The original non-iron cotton using a cross-linked resin finish was developed in 1969 by Dr Ruth Rogan Benerito, a scientist at the US Department of Agriculture. The initial results, though effective, left the fabric with a harsh handle and did not prove popular (Benerito, undated). The technology has become much more refined to the point where leading retailers such as Brooks Brothers can estimate that 70–80 per cent of their sales in women's woven shirts are in non-iron cotton. Joe Dixon, the company's vice president of production and technical design attributed its popularity to women's desire to look 'business smart' and the fact that 'women don't like the feel of starch, and they resent the higher cost of dry cleaning'. It would appear in the eyes of the retailer at least that environmental concerns are not high on their customers' shopping lists.

The Lenzing Group was presented with a Vienna Stock Exchange Award 2010 for its work on sustainable textiles (Lenzing Group, undated). In contrast with Brooks Brothers perception of the consumer, Peter Untersperger, Lenzing's CEO commented at the time of the award that, 'more and more customers in the textile and non-wovens industry demand ecologically sound and sustainable products'. One of the company's leading brands is Lyocell, a cellulose fibre made from wood pulp and marketed under the trade name Tencel. The wood pulp is converted into nanofibrils using nanotechnology so that the result is an exceptionally fine yarn that offers high moisture management with good tactile properties. Tencel absorbs excess liquid then quickly releases it into the atmosphere helping to keep the body at a comfortable temperature. This function carries an additional anti-bacterial benefit: because excess water is moved away from the skin bacteria have less opportunity to grow. The smooth surface of the yarn makes it pleasant to wear and gives the cloth a good handle and drape.

Smartfibre AG produces a lyocell fibre with enhanced health properties. SeaCell sees the lyocell fibre incorporate seaweed as an active ingredient. Seaweed has long been used in Chinese medicine and is recognized as offering protection for the skin and containing anti-inflammatory properties. There is what is termed an 'active exchange' between the fibre and the skin. The company describes how there is a movement of substances between fibre and skin with nutrients such as calcium, magnesium and vitamin E released by the body's natural moisture when a garment using the fibre is being worn.

45

Agave sisalana, or sisal, is most commonly associated with rope and twine making. The advent of cheap synthetic substitutes in the second half of the last century led to a fall in demand and production but this is now increasing as it has become popular in other applications such as carpets, automotive interiors and as an alternative to glass fibre in certain composites. The hardy agave produces some of the strongest and longest plant fibre with leaves measuring as much as 1 metre in length. Fibres are extracted from the length of the leaf, then extensively washed during decortication, using a rotating wheel set with blunt knives, before being sun-dried, brushed to separate and align the individual strands of fibre, graded and packed into bales. The plant is of particular benefit to countries with poor soil as it can be grown in arid conditions and needs little fertilizer or pesticides. On the negative side, the decortication process is water intensive and can cause pollution if used water is allowed to flow into the watercourses (Hamilton and Bensted-Smith, 1989). The biomass that remains after the fibres have been removed represents as much as 98 per cent of the plant, most of which is now flushed away as waste (Natural Fibres, 2009). This amounts to an estimated 15 million tonnes annually. The Common Fund for Commodities, UNIDO and the Tanzanian sisal industry have come together to fund the first commercial plant to use sisal residues in order to produce biogas, electricity, process heat and fertilizer. Indications are that 75 per cent of the energy produced in the plant could be distributed to rural homes with the remaining 25 per cent used in sisal processing.

While the market for sisal suffered because of the introduction of man-made fibres, hemp use as a fibre has struggled for a very different reason – marijuana. Although its use in textiles can be traced back to 8000 BC it is the association with hallucinogenic drugs that brought it to public attention in recent times. This is now changing as it has been shown that the plant used for fibres has less than 1 per cent of Tetrahydrocannabinol (THC), the active chemical in cannabis. This is far too little to be used as a recreational drug. The plant has the benefit of growing quickly, yielding more fibre per acre than any other crop (Earth Pledge, 2009). It renews the soil rather than draining it of its nutrients, can be grown without chemical spraying for pest control and is biodegradable.

Henry Ford once said that, 'The most environmentally friendly thing you can do for a car that burns gasoline is to make lighter bodies.' Ford's own experiments with composite alternatives to metal panels for cars included the use of hemp. Though there was enthusiasm about the results, the early composites did not make it into production, perhaps due in part to the US introduction of the Marihuana Tax Act of 1937 that effectively destroyed the hemp industry. Today, automotive industry-related R&D into more sustainable fabrics has undoubtedly been influenced by the European Union's End of Life Vehicles directive (End of Life Vehicles, 2010). This requires that all new vehicles should be 95 per cent recyclable by 2015. The challenge is to find ways to make composites for vehicles stronger and lighter, using them to replace heavier and less sustainable metal parts. The Ford Motor Company's Environmental Sustainability Report 2009–10 announced that: 'We are actively researching and developing renewable material applications that will reduce our overall use of petroleum products and improve our carbon footprint, while providing superior performance' (Ford Motor Company, 2010). In 2009 the company estimated that almost 300 parts in their European vehicles were derived from natural materials such as cotton, wood, flax, hemp, jute and natural rubber. They are now working on substituting up to 30 per cent of their glass fibre reinforcement with hemp and sisal for injection-moulded plastics.

Opinion is sharply divided as to whether bamboo is sustainable or not. The issue revolves around water. The fast growing plant can reach its full height in about three months, spreads easily with little intervention, pesticides, fertilizers or water needed to grow. It is biodegradable, naturally antimicrobial, thermodynamic and protects against ultraviolet (UV) rays. The issue begins with the processing and subsequent labelling of bamboo. One method is similar to that used in hemp and involves mechanical crushing to extract the fibres, with natural enzymes helping to further break them down. In the second, fibres are chemically broken down using lye, carbon disulphide and strong acids, then extruded through mechanical spinnerets. A further product, bamboo charcoal is a rayon bamboo fibre whereby the carbon from burnt bamboo is microencapsulated within the fibre to enhance the antimicrobial properties. The first method of production is a natural bamboo fibre while the second is bamboo rayon with a world of difference from an environmental point of view. There is a call for the labelling of rayon derived from bamboo to say just that, rather than a generic bamboo. The Competition Bureau of Canada has taken a stance on the matter and in January 2010 announced that 450,000 textile articles were being relabelled and over 250 website pages corrected (Competition Bureau of Canada, 2010). That is a staggering figure for a country with a population of just 33 million, half that of the UK.

There is a growing move to apply technological advances to nature to produce what are in effect techno-naturals. Du Pont's Applied Science Division has developed a way of turning cornstarch into fibres (Du Pont, 2010). Sorona is made from starch extracted from the kernel of corn and the company is also looking at ways of extracting starch from the stalk and other parts of the plant. The starch is put through a fermentation process where glucose is fed down pipes into a three-storey vat that contains genetically engineered organisms, water, vitamins and minerals. The resulting monomer is polymerized with a petroleum-based polymer and cut into pellet form that is converted into fibres. A key reason given by Du Pont for the growing interest in this area is the increasing cost of petroleum that had previously been less expensive than sugar.

The creators of Agent Orange and at the forefront of genetically modified (GM) seeds, Monsanto is no stranger to controversy. The company website declares that, 'While we share the name and history of a company that was founded in 1901, the Monsanto of today is focused on agriculture' (Monsanto, 2010). The company was the first to introduce GM cotton under the brand name Ingard in 1996, followed more recently by Bollgard II, with the promise that it would reduce the use of pesticides by 80 per cent compared with conventional crops. This in turn, it argues, cuts down on contamination of the waterways. Monsanto is currently developing new water use efficiency and nitrogen use efficiency traits aimed at further conservation of water and nitrogen.

47

NURTURE

> Others [architects and designers]... are convinced that the problems of high technology require a 'techno-fix, that is, the use of even more technology to solve the technology-based problems that we face on the planet (Papanek, 1995).

The demand for more environmentally friendly man-made fabrics is forcing the industry to take a long, hard look at itself. Every aspect of production is being carefully scrutinized with some strong regulatory and accreditation bodies performing a vital role in facilitating this process. The Oeko-Tex (2010) certification is the most universally recognized and accepted. To receive the Oeko-Tex® Standard 100 label, evidence must be provided of the successful testing of textile products for harmful substances. To ensure standards are maintained each certification is only valid for a 12-month period. Industries such as the European automotive are demanding fabrics that can be recycled in response to the EU's End of Life Vehicles directive. This has led to a change in attitude by textile manufacturers with environmental credentials being sought for products and listed on specification sheets alongside other fit for purpose requirements. Designers from a range of industries using textiles have also helped to oversee the evolution of more environmentally friendly synthetics. Design companies using textiles also play a significant role. We have seen this with the sportswear company Patagonia who played a key part in the development and introduction of recycled soda bottle fleece launched in 1993 (Patagonia, 2010). Furniture designer Herman Miller was a leader in the move towards Design for Disassembly

ABOVE | Figure 1.4.2 This laser sintering process from Freedom of Creation (FOC) offers the possibility to produce products using monomaterials with zero waste

and whose first formal environmental goal in 1991 was to achieve zero landfill (Miller, 2010). This dialogue between regulators, designers and manufacturers is resulting in much more than new fibre and fabric developments. What we are seeing is the emergence of new systems for design, manufacturing and consuming. These initiatives are not only more sustainable, but are producing some of the most exciting new developments to come from the textile industry in recent years.

Wellman International GmbH (Wellman International, 2009) uses post-consumer waste to make its polyester PET fibres. The PET bottles are collected and delivered to two recycling plants, one in The Netherlands and one in France. There they are first sorted then cleaned and chopped into small flakes before being transferred in bulk to the company's fibre plant in Ireland. Although the transfer between different processing plants is not ideal, the company estimates that the process has a carbon footprint almost four times smaller than a virgin polyester staple fibre. The fabric derived from the process can be used for applications ranging from hygiene to automotive interiors.

The Japanese textile manufacturer Teijin has taken the innovative step of involving the consumer directly in the production of its polyester PET fibres. Ecocircle is a closed-loop recycling system that takes polyester products at the end of their use and turns them into polyester PET fibres at its Matsuyama plant (Ecocircle, 2006). The company encourages consumers not only to buy products with the Ecocircle label, but also to bring them to a recycling bin at a participating store or post them to Teijin direct. Its website includes information in the form of an animation that encourages the consumer to engage with the idea that synthetic products can evoke their own sense of empathy and through the recycling process we may end up wearing that favourite soft toy we have outgrown as a shirt. A sense of engagement comes more readily with natural materials that people are more inclined to keep for longer, repairing and valuing much of the wear and patination that occurs. The Teijin message is an important one in getting users to think again about how they engage with and ultimately dispose of products that use man-made materials.

Teijin has developed a number of fibres using the process including Microft, Welkey and Cortico. All are high performance health-giving fibres that build on the company's expertise in highly engineered fibres. Microft is a soft handle, high-density microfibre with a water repellent finish used in sports and leisurewear. Wellkey is a hollow fibre specially designed to wick perspiration away from the body. Cortico is also intended for moisture control and uses a highly engineered triangular fibre containing small apertures on the surface to absorb perspiration more readily.

Knitted and woven structures form the cornerstone of the fabric construction industry, with composites, sandwich structures, non-wovens and braiding also well established. However, there are some new manufacturing technologies emerging that offer some exciting new possibilities for producing sustainable materials. Because these are in their infancy, often still in the research laboratories, their environmental credentials are largely speculative at this stage. As some of these are likely to occupy an important role in the future textile industry it is worth considering some of the possibilities they offer.

The process of laser sintering, or 3D printing as it is sometimes known, is where lasers are used to fuse fine particles of powdered nylon, plastic or metal to form strong three-dimensional objects. The benefit of the process is that whatever excess powder remains after the sintering can be reused so that there is zero waste. Generally regarded as an industrial design and prototyping process, the Dutch designers Freedom of Creation (2010) have explored the possibilities of the process in their textile-based work for fashion and accessories. While solid forms can take many hours to produce, the Freedom of Creation designs are not unlike modern day chain mail, intricately linked patterns but extremely lightweight. The garments and bags are seamless and can be of a mono-material. A further benefit is that small production runs are possible reducing the risk of unsold stock building up in warehouses or being wilfully destroyed as H&M were accused of doing in New York City in an article by the New York Times journalist Jim Dwyer (Dwyer, 2010).

Nanotechnology is manufacturing at a molecular level, smaller than 100 nanometres (nm) or 100 millionth of a millimetre. In essence this is the building of structures, including materials and coatings, atom by atom. The technology is at a relatively young stage of development although products are being launched in a number of areas including coatings for textiles.

The leading Swiss performance fabric manufacturer Schoeller was one of the first to bring a nanocoated textile to the market with its nanosphere coating (Schoeller Textiles, undated). This is an ultra-fine water repellent coating whose engineering was inspired by the lotus leaf, which has a bumpy surface texture and wax-like finish to repel water, taking dust and other dirt with it. The impact of the coating on textiles is referred to by the company as the 'self-cleaning effect'. Water and dirt are repelled, as is oil. Because of this function, nanosphere coated fabrics require less frequent and lower temperature washing.

However, it is the next stage of nanotechnology where things promise to get really exciting and we see the production of high strength carbon nanotubes. Nanotubes (Nanotechnology

Now, 2010), as the name suggests, are a tube shaped material made from carbon atoms bound together to create a stiff structure that forms the strongest existing bond. They are categorized by their structure, which can be single-walled, multi-walled or double-walled. Initial progress has been good, but scale and the cost of production still has to be overcome in order to make it a viable manufacturing process.

The Belgian company Nanocyl (2009) specializes in the production of carbon nanotubes. It uses a method of catalytic carbon vapour deposition because of its reliability and relatively large scale. In nano terms large scale means a measurement in centimetres. Its single-walled nanotubes have a single cylindrical graphite wall and come in the form of tubes that are capped at either end. The structure is visualized as a graphite layer just one atom thick, referred to as graphene, which has been rolled into a seamless cylinder. Single-walled nanotubes are typically 1nm in diameter but the length is much longer. They are pliable and can be twisted, bent and flattened without breaking. The nanotubes also display a unique set of electrical and mechanical properties giving them a wide range of applications including nanocomposites and nanosensors. The company is working in partnership with 3B-Fibreglass in order to develop CNT-Sized Glass Fibres for thermoplastic and structural composites. The intention is to combine the two technologies and joint performances so that the resulting fibres will be mechanically strong and electrically conductive.

ABOVE | Figure 1.4.3 Hybrid yarns such as Grado Zero Espace's Vectrasilk combines the performance of a synthetic with the handle of a natural fibre for maximum consumer appeal

51

ABOVE | Figure 1.4.4 Mambo proving that the environmental message can raise a smile while being effective

HYBRID

Hybrid materials see the bringing together of two or more different materials to make a new one with enhanced performance characteristics. In a fabric context, the term is usually used to refer to combinations of natural and synthetic or textile and non-textile. The benefit from a sustainable point of view is the temperature and frequency of washing that these new hybrids allow for, as well as their extended lifecycle. The negative is that they are not by their very nature monomaterials and with high performance elements often including hard to recycle metals or liquid crystal polymers (LCP) these are not in themselves ideal from a sustainable point of view. The aim in developing these materials is to combine the best of both worlds with some interesting results, aesthetically, in terms of performance and environmentally.

Technically, it is becoming easier and more cost-effective to combine the natural and synthetic as can be seen in advances such as Grado Zero Espace's Vectrasilk NT32 (GZE, undated). The yarn combines the high strength of the Vectra LCP with the handle and sheen of the silk (Ticona, 2010). An LCP is a highly crystalline, thermotropic (melt orienting) thermoplastic that has high thermal performance and chemical resistance. Vectra has been used for airbags on the Mars Pathfinder mission because of its high performance. In clothing applications where high performance is needed there is also a requirement for comfort. In developing a hybrid of LCP for performance and silk for comfort, the resulting material adds to the durability and longevity of the garment.

ABOVE | Figure 1.4.5 Fashion Technology: The Rip Curl Project saw UTS students explore a range of possible solutions to reusing neoprene. This example is from student Tara Savi and sees the material separated, layered, bonded and fringed. Photographer Carmen Lee Platt, Encapture Photography

In Switzerland, the Schoeller Spinning Group has developed a hybrid wool and stainless steel yarn for use in anti-static clothing. In the yarn, the Merino wool is randomly interspersed with less than 10 per cent Inox stainless steel fibre. The resulting yarn can be easily dyed, knitted and finished with treatments such as dirt repelling Teflon without interference with the shielding performance. The ability to combine capabilities such as these in a single yarn is relatively new. Less than a decade ago this would not have been possible without reducing, or even destroying the performance of one or both functionalities.

CONCLUSION

'This bag is made from trees that were blocking ocean views' (Mambo, www.mambo.com.au).

The bag is in fact made from brown paper that has been 'recycled and carbon positive and that'. The tongue-in-cheek advertising campaign is a reminder that the desire to produce and consume in a more sustainable way can force us to make some difficult decisions. The Norfolk Island Pines that line Sydney's Manly Beach (home of Mambo) are indeed majestic but they do also block the view. While no one would consider felling the Manly trees, environmental vandalism will continue until our behaviour changes.

Natural, man-made and hybrid combinations of fibres are making a concerted effort to improve their environmental impact. The scale of the problem is such that individually none of these developments will make the changes necessary to achieve a real difference. What is needed is to see these developments coupled with individual and collective responsibility and action.

In the introduction to this chapter we saw that the percentage of water withdrawal by Indian farmers for their agricultural needs is 90 per cent. In the section on Nature we saw how Monsanto is offering one solution to this in the form of GM crops. But there are other options, often happening at a local level and reliant on changes in farming practice and cooperation for success. One such initiative is the Bharati Integrated Rural Development Society (BIRDS). BIRDS is running the Andhra Pradesh Farmer Managed Groundwater System Project to help manage water use at a community level. Rainfall is measured and recorded and a water budget drawn up. Farmers voluntarily participate and an agreement is reached as to who should grow which crops. Account is taken of the water demands of each crop (with cotton one of the highest) and these have been reduced in order to enable less thirsty crops such as lentils and peanuts to be grown. The benefit of this is that it allows food to become a greater proportion of what is grown.

The iconic Australian sportswear company Rip Curl have launched a programme Project Resurrection to look at how they might reuse old wetsuits. The garments are collected and recycled to be used as the outer soles of footwear. In 2009, with Alison Gwilt, the author instigated Fashion Technology: The Rip Curl Project with year three students at the UTS Fashion and Textiles course. Students were asked to develop a series of fabric swatches, design and make a jacket using a combination of old wetsuits and fabric left over from the production process, which is a highly wasteful, 'cookie-cutter' approach. Their challenge was to make garments of the same if not greater value than the original. This proved a major challenge to the students, forcing them to rethink their approach to materials as well as their pattern cutting and assembly of the garment. A selection of the garments and fabrics were exhibited in a show that coincided with the Sydney Rosemount Fashion Week in April 2009.

These examples and the fibres and fabrics discussed in this chapter are part of an ongoing process of addressing how we make them, use and reuse fibres and fabrics. As our understanding of the environment and our impact on it continues to grow so does our response. What we do individually and collectively is our challenge and our responsibility.

53

REFERENCES

Benerito, R. R. (undated) http://web.mit.edu/invent/a-winners/a-benerito.html (accessed 15 June 2010)

Bharati Integrated Rural Development Society (BIRDS) (2006) 'Andhra Pradesh farmer managed groundwater system'. http://birdsorg.net (accessed 16 June 2010)

Competition Bureau of Canada (2010) www.bureaudelaconcurrence.gc.ca/eic/site/cb-bc.nsf/eng/03193.html (accessed 17 June 2010)

Cotton Incorporated Lifestyle Monitor™ (2010) www.cottoninc.com/lsmarticles/?articleID=35 (accessed 17 June 2010)

Du Pont (2010) www2.dupont.com/Sorona/en_US/ (accessed 17 June 2010)

Dwyer, J. (2010) 'A clothing clearance where more than just the prices have been slashed', *The New York Times*, 5 January 2010, pA16, NY edition

Earth Pledge (2009) www.earthpledge.org/sustainable-fibers.php (accessed 30 June 2010)

Ecocircle (2006) www.ecocircle.jp/index_e.html (accessed 17 June 2010)

End of Life Vehicles (ELV) (2010) http://ec.europa.eu/environment/waste/elv_index.htm (accessed 28 May 2010)

Ford Motor Company (2010) *Sustainability Report*, 2009. www.ford.com/microsites/sustainability-report-2009-10/environment-products-materials-sustainable (accessed 28 May 2010)

Freedom of Creation (FOC) (2010) www.freedomofcreation.com/ (accessed 27 May 2010)

Grimond, J. (2010) 'For want of a drink: Special report on water', *The Economist*, May 2010, p3

GZE (undated) www.gzespace.com/gzenew/index.php?lang=en (accessed 30 June 2010)

Hamilton, A. C. and Bensted-Smith, R. (eds) (1989) *Forest Conservation in the East Usambara Mountains, Tanzania*. International Union for Conservation of Nature and Natural Resource (IUCN), Gland (Switzerland) and Cambridge (UK)

Laser Sintering (undated) www.lasersintering.com/ (accessed 30 June 2010)

Lenzing Group (undated) www.lenzing.com (accessed 30 June 2010)

Marihuana Tax Act of 1937 (1937) en.wikipedia.org/wiki/Marihuana_Tax_Act_of_1937 (accessed 16 June 2010)

Miller, H. (2010) 'How we do it', *Environmental Advocacy*. www.hermanmiller.com/About-Us/Environmental-Advocacy/How-We-Do-It (accessed 16 June 2010)

Monsanto (2010) www.monsanto.com (accessed 30 June 2010)

Nanocyl (2009) www.nanocyl.com/ (accessed 30 June 2010)

Nanotechnology Now (2010) *Nanotubes and Buckyballs*. www.nanotech-now.com/nanotube-buckyball-sites.htm (accessed 30 June 2010)

Natural Fibres (2009) www.naturalfibres2009.org/en/stories/sisal.html (accessed 17 June 2010)

Oeko Tex (2010) www.oeko-tex.com (accessed 30 June 2010)

Papanek, V. (1995) *The Green Imperative: Ecology and Ethics in Design and Architecture*. Thames and Hudson, London, p25

Patagonia (2010) www.patagonia.com/web/us/patagonia.go?slc=en_ US&sct= US&assetid=2791 (accessed 30 June 2010)

Schoeller Textiles (undated) www.schoeller-textiles.com/en/technologies/nanosphere.html (accessed 30 June 2010)

Ticona (2010). *Vectra*. Ticona: Performance Driven Solutions. www.ticona.com/products/vectra (accessed 16 June 2010)

Wellman International (2009) www.wellman-intl.com/ (accessed 30 June 2010)

WELS (2010) Water Efficiency Labelling and Standards, www.waterrating.gov.au (accessed 30 June 2010)

MAKE
Chapter 2

MAKE | INTRODUCTION

This chapter focuses on approaches to sustainable design and the manufacture of fashion garments. As fashion designers become aware of environmentally friendly fibres and sustainable textile solutions, the real challenge for the designer and the production team is to find ways to engage with sustainable strategies within their design and production process. The ideas and thoughts discussed within this chapter aim to challenge the conventional notion of how fashion garments are currently created and instead argue for alternative processes for designing and making clothes. Specifically the chapter focuses on how, where and what sustainable strategies the designer and the broader production team can utilize in the making of fashion garments.

Fashion garments that are frequently found in landfill waste are typically designed '...on a linear, one-way cradle-to-grave model. Resources are extracted, shaped into products, sold, and eventually disposed of in a "grave" of some kind, usually landfill or incinerator' (McDonough and Braungart, 2002, p27). This cradle-to-grave model of producing fashion stems from the mid-19th century when a modern fashion industry emerged in Paris, and it continues remarkably unchanged to this day. It seems timely then to question the process of producing fashion garments and to seek alternative methods and solutions that can improve industry practices. However, the production of fashion is a complex system that involves and impacts upon many different resources, people and processes.

57

The people employed within the fashion industry make a vital contribution in the process of creation; the fashion system employs designers, buyers, pattern makers, machinists, knitters, textile designers, finishers and dyers, production managers and so on, and each brings specialist skills and knowledge. However, according to Papanek we should all reflect on our role in society and, while the complex sustainability issues and problems are often left to the professionals, the scientists and the activists, each individual should consider how he or she can create change at a local level (1995). Alison Gwilt explores the role of the fashion designer within the design and production process, and in particular questions what it is that designers do and why they do it. Gwilt argues that the fashion designer can be instrumental in changing the way that garments are designed and produced but there are some hurdles to overcome if we are to see designers really engaging with sustainable strategies. Meanwhile Holly McQuillan explores the strategy of zero-waste pattern-making as a technique for waste minimization, while referencing the way in which designers can engage with creativity through zero-waste strategies. McQuillan discusses the idea that designers need to appreciate the importance of taking risks in design practice if zero-waste strategies are to work effectively and the examples of her own design practice demonstrate this explicitly.

McDonough, W. and Braungart, M. (2002) *Cradle to Cradle: Remaking the Way We Make Things*. North Point Press, New York

Papanek, V. (1995) *The Green Imperative: Ecology and Ethics in Design and Architecture*. Thames and Hudson, London

ALISON GWILT

2.1 PRODUCING SUSTAINABLE FASHION:
The Points for Positive Intervention by the Fashion Designer

INTRODUCTION

This paper questions the process of fashion design and provides an argument for how fashion designers can engage with sustainable practices. By specifically discussing the fashion design process as applied in micro, small and medium sized fashion businesses,[1] the role and influence of the fashion designer can be mapped across the phases of design and production. As the phases within the fashion design process are identified, associated activities become a point of reference from which to consider how designers need to rethink their roles and behaviours in the context of sustainable design strategies.

An increasing number of designers are aware of their responsibility to engage with sustainable and ethical practices, but often feel unable to work within a sustainable framework (Centre for Sustainable Fashion, 2008). Moreover, Lawson (2006) suggested that there is a need to question the appropriateness of a conventional design process, if designers are to prepare for a changing future industry. In mapping the phases of production and the activities within the fashion design process it becomes possible to identify a conventional fashion design process, a critique of which, I believe, offers the opportunity for a holistic engagement with sustainable design principles. This holistic approach is essential if we are to see the fashion designer successfully integrating sustainable design strategies within the fashion design and production process.

THE PRODUCTION OF FASHION: A BRIEF HISTORY OF PROCESS

The global production of fashion involves numerous companies, manufacturers and retailers who develop products for specific sectors and market levels. Although the sectors of the fashion industry are broad (ranging from sportswear to lingerie) a company will typically operate within one sector while characteristically corresponding with womenswear, menswear or childrenswear (Stecker, 1996). A company will then generate one or more product lines, which fits precisely within an identified market level that is distinguishable by a defined set of criteria determined by factors such as price point, quality of material, manufacture and production numbers (Stecker, 1996). It is then the fashion designer's responsibility to create a new collection that fits within these criteria. The question is, how can a designer ensure that a product meets the defined market criteria and sustainability objectives at the same time?

Although the activities and duties of the fashion designer differ across all the levels and sectors of the fashion industry a conventional fashion design process does exist. This process emerged with the phenomenon of the fashion designer during the mid-19th century. While the evolution of an early fashion style can be traced to the French court of King Louis XIV in the late 17th century, the modern fashion industry materialized with the establishment of the couture business of Charles Fredrick Worth in the mid-1800s (Breward, 2003). Following political unrest, Paris emerged as a centre for luxury and style for affluent clients and new nobility. As mechanization increased the proliferation of products there was now a need for a conduit to act between the client, and the producers and merchants in the creation of fashionable clothes. Deemed the '…grandfather of French couture…' Worth opened the first couture house in 1858 and centred on the production of fashionable made-to-measure garments that provided his wealthy clientele with exclusive designs and quality craftsmanship (Breward, 2003, p29). Worth's clients were required to make an appointment whereupon he would assess their features and personal tastes, and would then design a gown that he considered suitable. Worth developed his own sketchbooks of design ideas and fabrics and so the fashion design process, as we know it, was beginning to be shaped into an intended practice that considered fabric, form and the body's proportions through graphical interpretations and ideas. Worth provided a unique service that placed a singular person, the fashion designer, in a position of responsibility for the entire process of garment design and manufacture (Wilson, 2003; Troy, 2003). In the role of fashion designer, Worth and his contemporaries capitalized on a cultural shift as the influence of fashion moved beyond the French court; the success of the early fashion designers '…was in their ability to read the implications of cultural and stylistic change and incorporate it into a characteristic and very well-promoted personal vision' (Breward, 2003, p23).

By the middle of the 19th century two systems of fashion production had materialized: the bespoke work of the tailor and couturier and the expansion of a ready-to-wear clothing industry that was in particular developing in the US and the UK. Beginning with the production of uniforms and menswear daywear garments, a ready-to-wear industry emerged in response to a need for stock supplies of ready-made clothing[2] (Wilson, 2003; Leopold, 1992). Although the manufacture of ready-to-wear clothing did not become factory produced until the beginning of the 20th century the impact of mechanization had a dramatic influence on the womenswear sector. Leopold defined this type of manufacture as, 'the investment in and coordination of labour and machines in a designated workplace for the purpose of increasing the productivity – and profitability – of manufacturing' (Leopold, 1992, p103). Even so, the fashion industry continued to rely on 'sweatshop' practices established in the 18th century, by employing poorly paid, mainly female, casualized staff (Wilson, 2003).

THE PHASES OF FASHION DESIGN AND PRODUCTION

The production of contemporary fashion clothing continues to rely on meeting market expectations within budget and manufacturing constraints and these factors have evolved to become the basis of the designer's brief. However, while fashion garments continue to be designed, produced and measured by economically driven factors, it is becoming increasingly imperative that we also measure the manufacture of a fashion product against its impact on the environment and society.

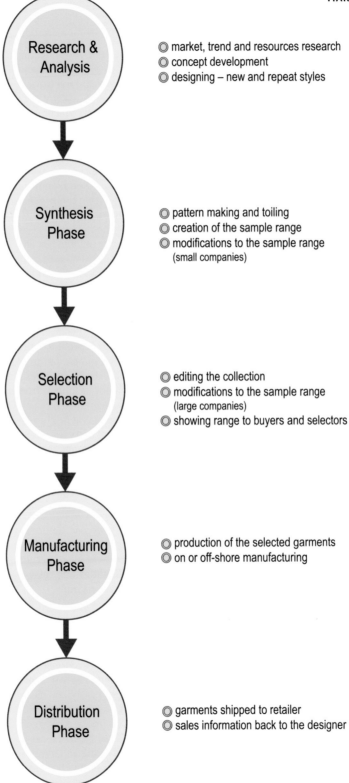

Research &
Analysis

◎ market, trend and resources research
◎ concept development
◎ designing – new and repeat styles

Synthesis
Phase

◎ pattern making and toiling
◎ creation of the sample range
◎ modifications to the sample range
   (small companies)

Selection
Phase

◎ editing the collection
◎ modifications to the sample range
   (large companies)
◎ showing range to buyers and selectors

Manufacturing
Phase

◎ production of the selected garments
◎ on or off-shore manufacturing

Distribution
Phase

◎ garments shipped to retailer
◎ sales information back to the designer

61

ABOVE | Figure 2.1.1 The five phases of fashion design and production. Sinha (2002)

The fashion design and production process involves a generic sequence of activities and phases that typically occur within all sectors of the fashion industry, which is well documented in many contemporary educational texts (Renfrew and Renfrew, 2009; Sorger and Udale, 2006; Jackson and Shaw, 2006; Jenkyn-Jones, 2002; Stecker, 1996). The process can be defined by five distinct phases: the research and analysis phase; synthesis phase; selection phase; manufacturing phase; and distribution phase (Sinha, 2002, p7). Within Sinha's model, Figure 2.1.1, the five phases can be further dissected into particular activities and tasks: duties that may be assigned to the fashion designer, or designated to another member of staff, department or other unit or facility within the supply chain. How these duties are designated depends upon the scale of the company. For example, as the size of the company increases the fashion designer's engagement within the process may diminish and his or her duties become confined to a set of well-defined tasks and activities (Sinha, 2002). This is observed in large-scale manufacturing where the designer is typically involved in the research and concept phase only and unlikely to engage or influence any development within the synthesis phase and beyond (Sinha, 2002). In contrast, within a micro, small or medium business the reach and influence of the designer is vastly increased since the designer's engagement extends across the entire design and production process. However, as identified in Sinha's study there is a commonality in the phases that typically occupy the fashion designer. Typically these centre on the research and analysis phase, and the synthesis phase and a distinct series of activities that commonly occur including: market and trends research; design research; designing and editing the collection; fabric selection; pattern-making and toiling; costings; and the production of the sample range (Renfrew and Renfrew, 2009; Sorger and Udale, 2006). These phases and activities (as described above) remain the areas where the designer has the most engagement or impact from which to influence positive change, notwithstanding variations in the scale of the company. In fact these two phases, and their activities, could be alternatively labelled as the fashion design phase, which mirror the process as depicted within fashion educational texts (Renfrew and Renfrew, 2009; Sorger and Udale, 2006; Jackson and Shaw, 2006; Jenkyn-Jones, 2002; Stecker, 1996). While the list of activities noted within the fashion design phase is not exhaustive it can be assumed that in a micro, small or medium business the fashion designer will be engaged in these activities, and that the influence of the designer extends across all the phases of design and production to some extent.

To recap, by analysing the phases of design and production we can determine that an archetypal fashion design process does exist. Furthermore we can establish where the fashion designer fits within this system of production and where his or her influence can reach. Through an analysis of the designer's tasks, responsibilities and relationships the points for positive sustainable interventions can then be mapped out.

THE FASHION DESIGNER'S ROLE

Understanding the role of the designer is fundamental to this paper. In broad terms, the fashion designer initiates and generates design ideas that become new fashion garments; in industrialized nations it is usually a professionally trained person who is expected to have a high level of design ability (Cross, 2006). The fashion designer must possess a range of skills, and these abilities fall into one of two areas, namely 'creative' ability and 'technical' ability. In order to demonstrate creative ability the designer should be able to generate an original design from a unique concept or be able to interpret trend information into a novel design idea. On the other hand, the designer should also possess the technical ability to recognize the capabilities of a fabric and have the capacity to follow a garment through the entire process of development and production (Jackson, 2006).

Importantly, the designer must also be able to communicate the new product to the manufacturer and the client (Cross, 2006). This description is typically communicated through drawings, and in the case of fashion design this may be achieved through the use of a stylized fashion drawing that provides a complete view of the garment, and a technical or trade sketch that provides the garment specifications and details. This is then followed by the production of a prototype / mock-up or *toile* that enables the fashion designer to see the garment in a 3D realization. This prototype allows the designer to trial the garment on a body, as a form of testing, and provides the opportunity to reflect upon the design, assess its appropriateness and conduct modifications by evaluating its aesthetics, ergonomics, production capabilities and its potential for success in the marketplace.

However, before a product is generated a design idea or proposal has to be initiated and as Cross suggests, 'the generation of design proposals is therefore the fundamental activity of designers, and that for which they become famous or infamous' (Cross, 2006, p16). While particular design ideas may be perceived as innovative or new, Cross argues that in the majority of cases design ideas are reincarnations of preceding designs. This may be a contentious view, but in fashion design this point is best signified in the formal description of garment types, e.g. a full skirt, a fitted blouse or a tailored jacket. Traditionally a designer will sketch early design ideas, '...thinking with a pencil...' (Cross, 2006, p16) is how the stage could best be described. While sketching ideas the designer will be balancing a number of criteria that include meeting the factors set out in the client or the fashion company's brief, technical or production constraints and the designer's own aesthetic values. Through sketching the designer engages in a non-verbal process to find a suitable design solution according to set criteria.

While the tasks and responsibilities of the designer can vary across different sectors of the fashion industry, the criteria within the design brief can also alter. When designing for a high street retailer, for example, it is not uncommon to repeat silhouettes or styles that have achieved significant sales success in previous seasons. Subtle changes will occur to refresh these ideas: a new fabric choice, or a new print colourway may be used. These criteria would then be outlined within the design brief. However, this contrasts greatly with the task of the fashion designer in the designer ready-to-wear, independent designer or couture market levels, where the designer will be expected to set new trends in fashion for the coming season. Furthermore, within the luxury and couture market levels the fashion designer's role may be further elevated to that of creative or artistic director. John Galliano, at Christian Dior is a designer, for instance, who may be regarded as the creative director of a couture brand. The role in this case becomes a centralized creative position that may encompass the responsibility for the artistic direction of the brand image amongst other creative responsibilities (Jackson, 2006).

THE FASHION DESIGN BRIEF

One of the first tasks for the fashion designer is to assimilate the set of criteria by which new designs can be measured as appropriate. The company, client or designer may nominate the criteria, however, it is these that form the basis of the designer's brief. In planning the fashion collection the criteria will later assist the designer with the process of elimination and refinement of individual garment design ideas (Seivewright, 2007; Stecker, 1996). The designer's brief will include developing design ideas that, for instance: meet the consumer's needs; meet market trends; represent the brand or label's vision; work in relation to the body; are designed for a specific market, occasion, season or function; work in relation to fabric selection. It is also necessary that garments can be produced within the budget and that the required resources are accessible (Stecker, 1996). This is a typical design brief that a fashion designer will refer to when developing the collection. However, there is a second set of factors that the designer needs to consider and this relates to the development of individual garment design ideas within the collection. The designer needs to consider the design elements within the collection and ensure that these remain appropriate to the requirements within the design brief. Seivewright (2007) provides a list of design elements that include: form; proportion and line; purpose; garment details; colour; fabric; ornamentation and print; themed references; contemporary trends; target market and level, and genres in fashion. However, if the already extensive range of criteria discussed here becomes the standard measure for the success of a design idea, then where and when do any environmental or ethical criteria come in to play?

Within the contemporary educational texts that map the fashion design process and indeed the criteria for designing fashion garments, there is no mention of a requirement to consider sustainable design principles within the creation of fashion garments (Renfrew and Renfrew, 2009; Seivewright, 2007; Sorger and Udale, 2006; Jackson and Shaw, 2006; Jenkyn-Jones, 2002; Stecker, 1996). While Renfrew and Renfrew do comment on working with ethics (in a publishers' note at the rear of the text) the comment has been provided by the publishers to raise awareness in the '...next generation of students, educators, and practitioners...' (Elvins and Goulder in Renfrew and Renfrew, 2009, p169). Clearly, the points raised here demonstrate that the methodology applied within the fashion design process needs to be readdressed to incorporate sustainable design principles, but why are contemporary fashion educational texts continuing to promote a traditional method of fashion design practice?

RESISTANCE OR INEXPERIENCE?

Designers employed within a fashion company often find that their role includes working to a design brief that follows a particular vernacular. For instance a designer will have to develop a product that is in line with the label's accepted aesthetic so that it meets the expectations of an identified consumer, can be manufactured by means that are known to the company, and so customary patterns of design and production are formed. These customary patterns can be witnessed in numerous published case studies of the designers in the fashion industry. However, designers that continue to work within an established system of design and production typically appear to resist the inclusion of sustainable strategies, as environmental and sustainable concerns are not routinely included within the designer's brief or form part of the fashion design process. This raises a question: can we attribute this lack of inclusion to an archaic fashion design process or to the failings of the people working within this system? By exploring the fashion design process in relation to production it becomes apparent that different structures can pose different problems.

Ian Griffiths (2000) presents a case study of the fashion design process utilized in the designer ready-to-wear market level. As a designer for the Italian MaxMara label, Griffiths worked within a design team that produced a number of fashion lines for the label.[3] Griffiths' case study equally positions the importance of the design team as a contributor in the fashion design process alongside a committee of selectors: the agents, the buyers and the merchandisers. Separated from the design team, the selectors decide upon the retail suitability of the designed garment when it is in its sample form. If accepted, the garment goes into production and then into the stores where the sales team report back to the design team on sales numbers and customer feedback. For the designer in this situation the fashion design process is driven by a shared belief in the product: between the designer and the customer, and the intermediaries in between. Griffiths notes that the Weekend range produced by the MaxMara group developed a line that uses '...classic inspiration... focuses on reinventing or modifying categories of garment which have recognizable generic features, such as aran sweaters, duffel coats or safari jackets' (Griffiths, 2000, p85). This observation reiterates the point made by Cross that many fashion designers develop design ideas that are reincarnations of preceding designs. A point to note here, which is shared by Sinha, is that the process of design and production incorporated in large-scale manufacturing utilizes a vertical supply chain, where one phase may remain separated from the next.

Each phase or stage of design and production will involve different people often working within separate sections of the company and/or supply chain, which are at times situated in different geographical locations. Within a vertical system, the fashion designer works from within a design team that is isolated, in many instances, from the production team who handle the pattern-making and sample-making stage and the buyers and merchandisers who manage the decisions regarding the selection of products that are to be placed within the stores. This vertical system places the fashion designer in one phase of the process, thereby excluding the opportunity to interact with other key contributors, or oversee the product (and its processes) from start to finish. In fact, in this system the division between the designers and makers has become 'a keystone of our technological society' that in turn relies upon the designer to resolve complex issues that are often associated with the use of new technologies, processes and systems (Lawson, 2006, p24). The vertical design and production process is in complete contrast to that traditionally utilized by the couturier and the designer located within a micro, small or medium business.

Outside of the realm of large-scale manufacturing the designer is a significant and central member of a team and is '...ultimately responsible for the initial design ideas, right through to overseeing first samples for selling' (Renfrew and Renfrew, 2009, p27). Through a complex relational system the designer works with a diverse set of people that include the fabric merchant, textile designer, the buyers and extending perhaps to others such as the public relations agents. In this process the designer is placed within a decision-making position, and will be apportioned the responsibility for any changes within the design process. This, according to Renfrew and Renfrew (2009), means that the ability to communicate, as well as being creative, is an important asset in any designer. UK fashion designer, Giles Deacon, directs his self-named company and employs nine full-time and four part-time staff, plus a number of students who assist as part of a work experience programme. As a designer, Deacon will develop design ideas that are then passed to the pattern makers or 'creative cutters' (Renfrew and Renfrew, 2009, p38). The pattern makers will develop the patterns utilizing a combination of draping and flat pattern-making techniques, '...they draw in cloth on the stand, working spontaneously putting things on and often create happy accidents that are made into toiles for the collection' (Renfrew and Renfrew, 2009, p38). Everything is photographed and recorded with samples produced and overseen by Deacon. He describes the relationships within the studio as very close; and as the label produces eight collections a year – four for the Giles label, and four for high street retailer, New Look – the small team of staff is inevitably engaged within the creative process. In small companies, such as Deacon's, the work ethic frequently creates an 'all hands on deck' attitude, particularly as lead times shorten in the studio's race to complete the upcoming season's collection. This close and creative relationship effectively forms a mutual arrangement to some degree. Deacon intimates that he works collaboratively on the collection with the print designers, the knit designers and the pattern cutters. In addition he also conceives small design projects for the work experience students and design assistants, the results of which may or may not influence the new styles for the season. It appears that all the commonalities of the fashion design process are there: developing the concept for the collection; sketching design ideas; pattern-making and toiling; and producing the sample range. However, it is difficult to ascribe certain activities to the designer alone. Despite the creative process consciously or unconsciously becoming a collaborative effort, ultimately the collection has to remain the creation of the individual fashion designer: 'The interpretations of new styles have to look like they come from Giles' (Renfrew and Renfrew, 2009, p38).

POSITIVELY INTERVENING IN FASHION DESIGN PRACTICE
Aside from technical and educational texts that record the fashion design process, existing studies typically focus on the case study of eminent designers; or centre on the reporting of the

creative artefact from a particular designer (Sinha, 2002; Griffiths, 2000). Any analysis or critical study of the activities performed by the fashion designer appears to be insufficient; Kawamura goes as far as to suggest that, 'None of the writers discuss the role that designers or creators of fashion play in producing fashion...' (Kawamura, 2005, p58) and it follows that the area requires further exploration, particularly when traditional methods need challenging.

The approach to the creation of fashion garments synonymous with Deacon's label typifies the practices found within many micro, small or medium businesses. As an organizational system this fashion design process – from sketch to realized sample garment – places the designer within an influential position. Sinha supports the claims, commenting:

> The role of making in all of the companies fell to the pattern maker and the sample maker; the fashion designer took on a managerial role. All designers interviewed felt that the responsibility of overseeing sample-making was akin to maintaining their design integrity: the sample-making teams interviewed felt very responsible for making according to the designer's requirements (Sinha, 2000, p31).

The designer within a micro or small business such as Deacon's could be considered to be a hierarchical role, but in truth is perhaps best described as a centralized one. The designer works in a complex relational system that provides the opportunity to engage with a wide variety of skilled practitioners and companies in the creation of fashion clothing. From the centralized position the designer can influence and impart new (sustainable) information that aids in the design and production of 'better' garments.

From the observations discussed within this paper, it could be argued that the fashion design process, as it is commonly understood, should be challenged. If we are to acknowledge that a traditional model of fashion design practice exists then we need to acknowledge that there are conventional activities that engage the fashion designer. It is the notion of this conventional behaviour that needs challenging. While the fashion designer may be receptive to a break with conventional approaches if he or she is not positively exposed to the principles of sustainable design then the problem of poor engagement will continue. If fashion designers do not understand what sustainable design strategies are, how to engage with them and the possibilities that they offer then they are unlikely to alter their fashion design process. It is imperative that the contemporary fashion designer sees sustainable strategies in terms of the opportunities for innovation.

67

ENGAGING WITH SUSTAINABILITY

Efforts to engage designers (of all disciplines) in meeting the sustainable challenge are evident in the many publications about design, but which often depressingly reveal a typical 'business as usual' philosophy (Fuad-Luke, 2004). While eco or green products have begun to appear within the pages of design magazines more often than not these articles seem peripheral; within a popular fashion magazine the reader may see a one or two page feature of inspiring eco-fashion garments but the majority of fashion spreads in the magazine are typically dedicated to the conventional types of fashion. That is, garments that do not claim to or appear to address any environmental, social or ethical issues. Therefore, it could be suggested that neither the fashion designer nor the consumer is being regularly exposed to or educated about how fashion garments can be created and used in accordance with sustainable criteria in any meaningful sense.

To improve on the current piecemeal approach to sustainable issues in fashion design and to encourage the fashion designer to engage with sustainable design practice, I would argue that there has to be an appreciation of three key problem areas.

LEFT | Figure 2.1.2 Localized skills working together to create slow luxury fashion using upcycled waste materials and environmental print methods. 'Journey of an ornate sleeve' produced by illustrator Zoe Sadokierski, screen-printer Steve Woods, embroiderer Helen Parsons with fashion designer Alison Gwilt. Silversalt Photography

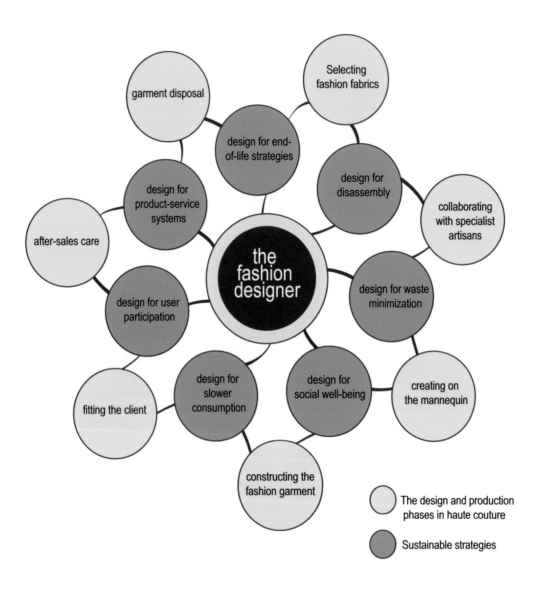

ABOVE | Figure 2.1.3 Linking sustainable strategies with the process of design and production. The model is indicative of the range of sustainable design strategies that can be linked to activities within the design and production process. Based upon the methodology applied in haute couture

Understanding Sustainable Design Strategies

Firstly, there is a need to assist the fashion designer in understanding sustainable design strategies, what they are and how they can be integrated within the system of designing and making fashion products. Fashion designers have tended to view sustainability as an afterthought to their design practice and so the integration of sustainable strategies within the fashion design process is not typically considered. Computerized tools, such as the Ecometrics calculator and others, aim to reduce the negative environmental impacts associated with the production and use of a garment. These tools can assist the designer during the production phase but while useful for quick responses that are required in large organizations, the use of these impact calculators may further lead a novice or unsuspecting designer to consider sustainability as an add-on rather than an integral part of the fashion design process (Black et al, 2009). Moreover, this solution-focused approach does not challenge or encourage designers to seek alternative strategies for designing and making clothes at the concept or research stage; these tools almost distract from the fact that the current paradigm does not encourage the integration of sustainable strategies within the fashion design process. Sustainable design strategies offer a structure or method through which the fashion designer can engage with sustainable design principles. A designer can choose to embody one or more strategies across the entire production process, for instance incorporating design for disassembly for easier product recycling, or durability in design or a zero-waste approach to material use in production. Advocates such as Jonathan Chapman and Nick Gant (2007), Sandy Black (2008) and others all recognize that the strategies currently being explored within design practice are not finite; there is still much progress to be made particularly in developing strategies that are all encompassing in their contributions to the fashion discipline.

Link Sustainable Strategies With the Fashion Design and Production Process

Secondly, fashion designers need to visualize sustainable design strategies integrated within their own design practice; this means that the designer should link sustainable strategies with the activities within the fashion design and production process. Lawson (2006) suggests that a designer should develop parallel lines of thought, a method that enables the designer to work out different aspects of the design at the same time. If the concept of parallel lines of thought were applied within the fashion design and production process a designer would be able to design, plan and create a new garment design in tandem with the integration of sustainable strategies. It is through these conversations that the fashion designer can begin to identify areas within the current paradigm for positive intervention.

In the example in Figure 2.1.3 Lawson's parallel line of thought was applied in a mapping of the step-by-step process of the design and production methodology applied in haute couture. The model suggests, for example, that a designer in haute couture can collaboratively work with the local artisans to creatively employ the strategy of design for disassembly by careful mono-material and ornamentation choices; or the designer who integrates design with the technique of draping on the mannequin can employ a zero-waste approach as a strategy for waste minimization; or the designer can apply high-end, durable construction finishes during garment manufacture and educate the wearer on garment care as a strategy of designing for slower consumption; and so on (Gwilt, 2009). In this context, the designer is situated within a central role, rather similar to the case study of Deacon discussed earlier within this paper. From a centralized position the designer can take a holistic approach to the lifecycle of fashion garments, and apply a single or a multitude of strategies across the entire design and production process.

While a perfect model for creating sustainable products does not exist as yet, the current objective must be to minimize negative and maximize positive impacts. Through a mapping exercise of the design and production process, broken down further into a sequence of activities, it becomes possible to think through the principles of a particular sustainable strategy while planning typical design tasks. As the fashion designer prepares to act out or engage in an

activity he or she can apply Lawson's method of parallel lines of thought – planning the task and applying the sustainable strategy – before taking action. With a good understanding of sustainable design strategies it becomes possible to select the most beneficial strategy for a particular point of intervention for any process of design and production: the designer needs to think and plan, then act.

Apply Lifecycle Thinking to the Fashion Design Brief

Thirdly, there must be a revision of the design brief to which a fashion designer works. Currently, few designers work from a brief that considers a lifecycle approach to design. However, there is great scope for applying lifecycle thinking within the fashion design brief to include, for instance, end-of-life strategies. Black et al (2009) argued that the common industry perception is that once a garment is in the hands of the consumer then it is no longer the responsibility of the designer. While a fashion designer may select a fabric for a garment based upon its performance and aesthetic qualities, that choice may vary if there is a consideration for what the wearer should do with the garment once they have ceased to wear it. A lifecycle approach would suggest that whether with an expensive jacket or a cheap fashion top, the designer should be questioning how consumers engage with a garment that they have designed. Research on the

ABOVE | Figure 2.1.4 Here Liu uses a mono-material approach to garment production, and a zero-waste pattern-making process

RIGHT | Figure 2.1.5 Chanin's label produces limited-edition garments that are handmade by local artisans living in the Florence region of Alabama. Using a combination of organic, recycled and new materials, the garments exemplify local traditional craft techniques

engagement between consumer and garment can critically inform the performance of a product in relation to the customer, company and sustainability criterion. Ultimately the designer needs to accept that the design brief must extend beyond the economically driven conventional criteria to include criteria that will meet the needs of the environment and society.

EMBRACING SUSTAINABILITY

It is not until we look at and critique the way that we teach, learn and use a process for fashion design and production that an alternative method of practice for fashion designers and makers will emerge. Lawson (2006) asks whether the disconnection of designing from making encourages better design? The fashion designer in a micro, small or medium size business does not work in isolation, as the creation of the fashion garment is a collective process (Kawamura, 2005). Therefore any opinion on the role of the fashion designer needs to be considered in relation to connections with other staff and units engaged within the fashion design process. The collective process provides the designer with a mechanism for positive change, and for the real integration of sustainable strategies within the fashion design and production process. Moreover, this relational design thinking extends the fashion design process to include a relationship with the consumer, who it is hoped will use, care for and dispose of the garment in a sustainable and responsible manner. Indeed the true challenge is not to design and produce sustainable garments but to encourage behavioural change within our society. The ideal scenario is one wherein designers drive design for sustainability through their approach to products, services and systems, which then advocates widespread change in the fashion production process and in the public's attitude to the consumption and use of fashion garments.

NOTES

1. The definition of a micro, small or medium fashion business differs according to the country of origin. In this case the model applied is based upon that defined by the Centre for Fashion Enterprise (2008): micro business – owner/designer; small business – 4+ employees; medium business – 13–15+ employees.

2. Although the industry's origins can be traced back further to the late 17th century with the production of large quantities of military and naval garments (Wilson, 2003; Leopold, 1992).

3. In this instance Griffiths describes and compares the creation of a jacket developed for two different seasons.

REFERENCES

Black, S. (2008) *Eco-chic: The Fashion Paradox*. Black Dog Publishing, London

Black, S., Eckert, C. and Eskandarypur, F. (2009) 'The development and positioning of the Considerate Design Tool in the fashion and textile sector', *Proceedings of the Sustainable Innovation Conference, Falmouth, UK*

Breward, C. (2003) *Fashion*. Oxford University Press, Oxford

Centre for Fashion Enterprise (2008) *The UK Designer Fashion Economy: Value Relationships – Identifying Barriers and Creating Opportunities for Business Growth*. NESTA, London

Centre for Sustainable Fashion (2008) *Fashion and Sustainability: A Snapshot Analysis*. London College of Fashion, London

Chapman, J. and Gant, N. (eds) (2007) *Designers, Visionaries and Other Stories: A Collection of Sustainable Design Essays*. Earthscan, London

Cross, N. (2006) *Designerly Ways of Knowing*. Springer, London

Fuad-Luke, A. (2004) *The Eco-Design Handbook: A Complete Sourcebook for the Home and Office*. Thames and Hudson, London

Griffiths, I. (2000) 'The invisible man', in I. Griffiths and N. White (eds), *The Fashion Business: Theory, Practice, Image*. Berg, Oxford, pp69–90

Gwilt, A. (2009) 'Generating sustainable fashion: Opportunities, innovation and the creative fashion designer', in E. Rouse, *Fashion and Well-Being* CLTAD / IFFTI, London, pp439–452

Jackson, T. (2006) 'Fashion design', in T. Jackson and D. Shaw (eds), *The Fashion Handbook*. Routledge, Oxford

Jackson, T. and Shaw, D. (eds) (2006) *The Fashion Handbook*. Routledge, Oxford

Jenkyn-Jones, S. (2002) *Fashion Design*. Laurence King Publishing, London

Kawamura, Y. (2005) *Fashion-ology*. Berg, Oxford, New York

Lawson, B. (2006) *How Designers Think: The Design Process Demystified* (4th ed.) Architectural Press, Oxford

Leopold, E. (1992) 'The manufacture of the fashion system', in J. Ash and E. Wilson (eds), *Chic Thrills: A Fashion Reader*. Pandora Press, London, pp101–117

Renfrew, E. and Renfrew, C. (2009) *Basics Fashion Design 04: Developing a Collection*. AVA Publishing SA, Lausanne

Seivewright, S. (2007) *Basic Fashion Design: Research and Design*. AVA Publishing SA, Lausanne

Sinha, P. (2000) 'The role of design through making across market levels in the UK fashion industry', *The Design Journal*, vol 3, no 3, pp26–44

Sinha, P. (2002) 'Creativity in fashion', *Journal of Textile and Apparel, Technology and Management*, vol 2, no 1V

Sorger, R. and Udale, J. (2006) *The Fundamentals of Fashion Design*. AVA Publishing SA, Lausanne

Stecker, P. (1996) *Fashion Design Manual*. Macmillan Education Australia, Melbourne

Troy, N.J. (2003) *Couture Culture: A Study in Modern Art and Fashion*. MIT Press, Cambridge, MA

Wilson, E. (2003) *Adorned in Dreams: Fashion and Modernity*. I. B.Tauris, London

73

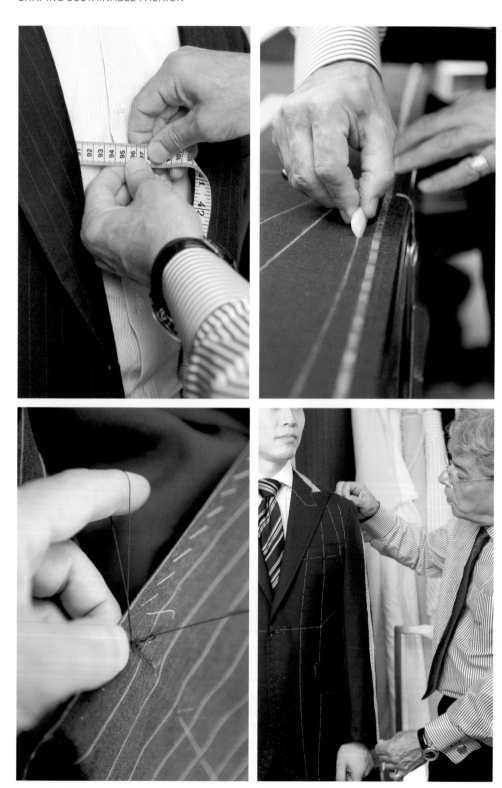

2.2 CASE STUDY
THE TAILOR'S CRAFT

It is commonly expected that the everyday business of most fashion companies relies directly on the production of a garment. Garments, which may have been designed and manufactured in different locations, are typically shipped to a retailer who then sells the items to an identified but anonymous consumer. While for the fashion designer in this chain there may be no direct relationship with the people who make, sell, or buy the garment, the focus for all of these people is on the continual creation of new products. This cycle of making and buying fashion clothes increases in speed as clothes becomes cheaper and more accessible. In many cases this encourages over-consumption and creates excessive amounts of textile waste that is generally dumped in landfill sites as consumers discard their unwanted clothes. But how can textile waste in landfill be reduced if we continue to make and sell new products? As a strategy for sustainable production and consumption the product-service system applied within the tailoring sector can provide a positive case study that could be an inspiration for contemporary fashion businesses.

The service supplied by bespoke tailors like Bijan Sheikhlary provides a client with a garment that has been individually handcrafted using the highest level of skills and materials. The design for a bespoke jacket emerges during discussions between tailor and client at the first fitting session while the tailor or his apprentice accurately takes measurements. The client will then select the garment fabric and suggest aesthetic details and personal needs, such as how many pockets are required and where they should be placed on the jacket. Throughout various stages of the jacket's construction the client will return for further fitting sessions to discuss and assess the jacket in terms of comfort and fit; aesthetic design; and purpose and function. The hand-sewn construction methods and carefully selected materials ensure that the jacket is durable while the use of excess seam allowance (inlays) prepares the jacket for later alterations if the client's figure or personal aesthetic tastes change. The clients are loyal and they continue to return to the tailor for services such as garment alterations and repairs, and for additional products to complement previous purchases. For the bespoke tailor the relationship with his client is paramount to the success of his business.

The lifecycle of a fashion garment can be extended through the use of a product-service system combination, which places a dual focus on both the product and a service rather than concentrating on the product in its entirety. A product-service model allows the fashion consumer to gain both a garment and a service through a number of different approaches. A garment can, for example, be complemented with a repair service, or an alteration or remodelling service, or the producer may provide a take-back scheme that allows consumers to hand the item back when they have ceased to wear it. These product-service models provide the possibility for new niche business opportunities that could revive ailing businesses, especially if the interest in a slow fashion movement gains impetus. As we seek to reduce the overproduction of fashion garments diverse business opportunities can flourish: including repair and alteration services; fashion leasing services; and remodelling services that can positively impact on the amount of textile waste being generated through excessive consumption.

LEFT | Figure 2.2.1 Stages of a bespoke jacket. Photographer Nick Bassett

OVER | Figure 2.2.2 Bijan Sheikhlary at work. Photographer Nick Bassett

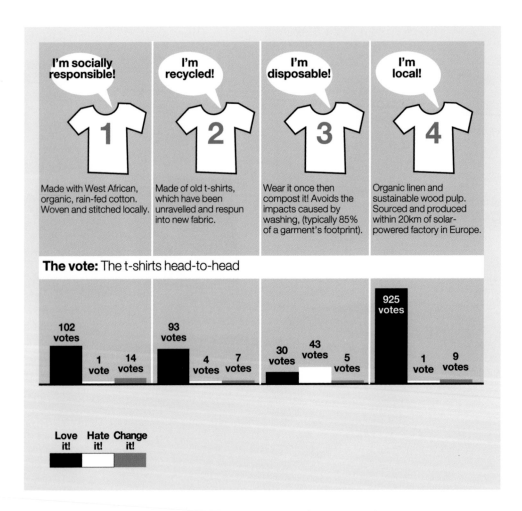

2.3 CASE STUDY
THE PERFECT SOLUTION

For all fashion producers the ultimate goal should be to achieve a 100 per cent sustainable garment but to many this ideal often seems impossible to realize. Jonathan Chapman and Nick Gant believe that it is more helpful if we think of 'degrees of sustainability' that can provide us with the opportunity to think about how improvements can be made while being creative and experimental in our approach (Chapman and Gant, 2007, p8).

For Better thinking Ltd, a creative and strategic design and consultancy company based in London, the redevelopment of the classic T-shirt provided the opportunity to expose the challenges in engaging in sustainable and responsible practices. 'Wouldn't it be nice, we thought, if there was a product that was as ethical as can be at every turn? That wasn't only less bad, but was absolutely the best it could be?' (Better thinking, 2007). The iconic T-shirt, chosen as a reflection of global consumerism, showcased real problems and solutions to both producers and consumers. For Mike Betts and Mark Holt at Better thinking, the purpose of the Perfect T-shirt project was to produce the most responsible garment possible and to share this newfound knowledge with the broader community. As the project traced the development of the garment from its inception through to the final sample product, the accompanying website documented the problems that were encountered and the solutions that were employed.

The project began in 2005, when Better thinking challenged the international community to specify the essential requirements needed for a perfect (sustainable) T-shirt. Having provided four different scenarios, the company asked the public to cast its vote on the website that then led to one produced garment. The T-shirt's journey began with a careful evaluation of material sourcing and garment manufacture. As the project progressed, experiments, challenges and discussions with suppliers and experts from materials, sustainability and garment manufacturing, were all aired on the website blog. The final product utilized organic cotton grown in Peru with its natural colour maintained, and was manufactured with energy from renewable resources by Fairtrade registered farmers and factory workers.

In 2008 the Luxury Redefined T-shirt was launched in retail stores as a result of a partnership between the UK knitwear company John Smedley and Better thinking Ltd. Now commercially produced in Derbyshire using organic, undyed, unbleached, fair trade Peruvian cotton, the luxury product is sold with a brochure that explains the journey of the T-shirt to the consumer, making the process of design and production transparent. The Perfect T-shirt project has received much acclaim, appearing in international exhibitions and has become a case study within academic programmes in the UK and internationally. The project demonstrates the real advantages of pursuing the ultimate goal and the benefits of reconceptualizing how we design and make fashion garments. The Perfect T-shirt project clearly demonstrated the potential of changing existing practices.

....

Better thinking (2007) 'Why is an ethical branding consultancy researching, designing and manufacturing a T-shirt?' www.betterthinking.co.uk/perfect/about.htm (accessed 28 June 2010)

Chapman, J. and Gant, N. (eds) (2007) *Designers, Visionaries and Other Stories: A Collection of Sustainable Design Essays*. Earthscan, London

OVER | Figure 2.3.2 Luxury Redefined, from UK knitwear company John Smedley/Better thinking Ltd

HOLLY MCQUILLAN

2.4 ZERO-WASTE DESIGN PRACTICE:
Strategies and Risk Taking for Garment Design

INTRODUCTION: WHY CHANGE?

Prior to recent advances in technology – most notably the internet – social and industrial uptake of changes in clothing trends occurred more slowly as information took longer to travel across the globe. In the contemporary world the combination of cheap production of textiles and apparel, and the rapid distribution of fashion trends have led to an industry that can respond rapidly to changing social mood.

Once styles are gleaned from the internet and processed by the design room, manufacturing advances mean that fast fashion firms such as Zara can take as little as 14–21 days from design room to retail floor (Tokatli, 2007; Tiplady, 2006) – enticing the consumer to purchase more, and more often, in order to remain up to date. However, the production and consumption of clothing has not always been so rapid, nor so seemingly effortless. Prior to the industrial revolution, clothing and cloth were expensive and time-consuming products to make – and as a result fabric use and pattern forms were carefully manipulated to use as much of it as possible – often resulting in close to 100 per cent yield (Burnham, 1973).

In our globalized contemporary world, information, raw materials and consumer goods are so readily accessible we often overlook the environmental impact of what we design, source, make or consume. The success of such convenience is clearly seen in fast food. Fast food is successful because it removes the need to decide on, purchase for and then prepare a home cooked meal – it is convenient and cheap. However, the impact of this food production and consumption model on the environment and human health is not reflected in the price at the counter. The same can be said for fast fashion – the advantages of this model of clothing production and consumption are convenience and low cost – but like food, it also carries enormous environmental and social impacts. In addition to the carbon generated through transportation, the huge volume of water and toxic chemicals used to grow and process textiles, and the use of poorly paid workers, it is also standard for garment producers to expect to waste approximately 15 per cent of the cloth needed to produce an adult sized garment (Rissanen, 2005; Feyerabend, 2004; Abernathy, 1999; Cooklin, 1997), resulting in a loss of profits for the manufacturer and generating landfill waste. When added to the rapid and insatiable desire for new fashion products that drives the fashion cycle, this contributes 30kg of textile waste per person in the UK every year (Santi, 2008).

Both the production and consumption of clothing needs to slow down to speeds that are sustainable in the long term. This paper examines my own design practice and that of others in the field of zero-waste garment design to offer alternative ways of approaching the making of clothes. What can borrowing from an ancient model for pattern-making and design offer in the attempt to slow a world so intent on having more, faster. What additional tools can zero-waste clothing design provide?

CERTAINTY AND RISK

Humans have been attempting to control the natural world for millennia. We like to be able to predict outcomes, to minimize risk and uncertainty, and we love to design processes and products in which we exert control over our environment and the organisms in it, giving minimal respect to the reciprocal relationship that living on the same planet must involve. As a result, the majority of these processes and products operate one way – we take good from the environment and give back our junk. The negative impact of practices used in the fashion industry can be most clearly seen in the fast fashion system. An email I received from a good friend who was working for a fast fashion producer in London highlights many of these. She wrote of the design process in the company she worked for:

> I don't actually design them. But, in the loosest sense of design, I 'adjust'. Are you laughing??? I do most days. I correct appalling fit, I decided on length / print / colourways. I rip out a Lacroix skirt (out of Vogue) that I love with loads of lace and send it out to the factory with a line drawing and basic spec, cross my fingers and hope that something nice comes back. It's fast, and crazy-busy. That particular skirt was lovely but came back at £5-6.00, so out of price range for a lot of our clients (Anon., 2008).

As a design process what is described demonstrates little risk taking or creativity, but is commonplace in fast, mass or mid-market fashion. The projected success of the 'Lacroix' inspired skirt relies on the already established success of the original and is therefore a safe investment – without risk and completely under control. The account above is backed by Tiplady who writes that 'unlike the retailers of haute couture and ready-to-wear, fast fashion retailers do not directly invest in design but instead are inspired by the most attractive and promising trends spotted at fashion shows and by cues taken from mainstream consumers' (Tiplady, 2006). Interestingly, Tiplady then argues that by producing fashion garments so rapidly in response to existing products, 'they also replace values such as exclusivity, glamour, originality, luxury and lifestyle, which were once "the fulcrum of ...fashion", with the values of trend, "massclusivity" and "planned spontaneity"' (Tiplady, 2006). But how much does a design process that results in a £5 'Lacroix' skirt being too expensive contribute to fashion consumption? In the fashion industry we take nutrients, sunlight, water, soil and plants to grow fibres such as cotton, and in return we give back a design, production and consumption cycle that contributes to global warming, social injustice and environmental degradation. As the certainty we had in our ability to continue to do this indefinitely begins to waver, we must consider what can be done.

The natural reaction as a designer to such a change is to attempt to regain some certainty in a new product, process or way of life – to believe that design can save the world. John Wood wrote that 'our egotistical attitude to innovation' (Chapman and Gant, 2007, p104) is more or less responsible for the environmental and economic mess we are currently in. It is frightening that the vast majority of so-called innovative and creative design results in products for which innovation is limited to aesthetics and a general desire for new things, which ignores issues of the environment and depleting resources. Our arrogance in our ability to control the world we live in has led us to this precarious state, so consider for a moment that control may not be the answer – instead we as designers need an openness to adaptation and to adopt a more holistic approach to design.

In this uncertain world it makes sense to use design processes that mimic the adaptive natural world. Generally, clothing collections are designed with a specific end product in mind, but can clothing be designed in a different way that better reflects the unpredictable world we live in and achieve this in a sustainable way, whilst still producing appealing fashion items? The general process from design to production follows design, pattern-making, construction and production. The separation and hierarchy of these processes has led to a cut and sew fashion system that is extraordinarily wasteful – the act of garment cutting ends in an average of only 85 per cent of effective textile use, leaving the other 15 per cent on the cutting room floor (Rissanen, 2005; Feyerabend, 2004; Abernathy, 1999; Cooklin, 1997). From these figures Timo Rissanen (Hethorn and Ulasewicz, 2008, p187) estimates that at least 100,000 tonnes of fabric is wasted in making clothes for the UK every year. This waste has traditionally been seen as a production problem – designers draw the garments, pattern makers make the patterns and when the pieces are laid out for cutting there is a marker made to minimize waste – but it is this hierarchical system that leads to so much of the fabric being wasted. However, this waste could be minimized if design and production were to become more integrated – and closer to the processes of nature. Wood (Chapman and Gant, 2007, p111) writes that the use of resources in a holistic production-to-consumption system would need to become a zero-waste system – as in nature. Indeed, if the fashion process were fully integrated it becomes possible to reduce the waste figure to zero, but only through a readjustment of acceptable levels of calculated risk and uncertainty within the design process, presenting a major hurdle for many designers.

In teaching my first year design students we encourage them to embrace uncertainty – to be OK with the unknown – yet few feel comfortable doing this. Within fashion design there are few such examples to show students to inspire them to take risks. Many fashion designers merely regurgitate past styles and follow the same well worn path from idea, to production, to retail and eventually to waste. One designer who deviates from this is Julian Roberts. His process does not follow the usual rules or order of design to pattern-making, his is a process where the design is the pattern-making is the cutting, resulting in garments that defy many of the norms of garment design, shape and form. His creations only reveal their form once on the human body – lending their freshness to an unpredictable and integrated design and pattern-making process that is developed from a range of rules he established himself (Roberts, 2010). He has called one of his processes 'subtraction cutting' as the final shapes are determined by what is removed and how the body travels through the space created. He has taken this method and applied it to a fashion system that designs for individuals, making personal that which is usually a slick sort of anonymity. This form of consumption encourages true ownership and attachment to clothing beyond the throwaway fashion cycle that is so dominant today.

Timo Rissanen is another designer engaging with new methods of fashion production. He identifies the segregation of the pattern-making and design process as what inhibits further evolution of the fashion system. Rissanen (Hethorn and Ulasewicz, 2008, p184) has developed a process of zero-waste garment design whereby his 'Jigsaw puzzle' method results in all pieces of the garment pattern being utilized in the finished design. To achieve this requires that he design both two-dimensionally – to achieve a jigsaw piece effect on the cloth with zero waste – while also considering the impact this has on the three-dimensional outcome of the garment. This method of fashion design relies on a certain degree of ease with unpredictability as there is – at least initially – informed guesswork involved. He writes 'if designers were open to some degree of trust in such unpredictability' (Hethorn and Ulasewicz, 2008, p202) then his puzzle piece method of making clothing would be more readily adopted.

Luis Eduardo Boza writes of the possibilities when combining two seemingly opposing strategies – computer numerical controlled machinery with handcrafted processes – for interior design. His students were encouraged to exploit unexpected events in the production process and to use the 'mistakes' the machinery – or their programming of it – made and 'as a result, the design/fabrication/assembly process expanded and exploited the findings from the intended and unintended discoveries. Ultimately, this process was informed through a physical contact with the material itself' (Boza, 2006). This process of risk and uncertainty with subsequent sensitive reaction can lead to unexpected positive outcomes in response to material, form and environment. The advantages of risk taking are not only shown in designers operating on the edges of traditional design practice. Nigel Cross states in his 1999 examination of design practice that 'design is opportunistic, and so the path of exploration cannot be predicted in advance' (Cross, 1999). He writes that designers have to recognize the risk inherent in good design practice, and be able to work in a creative environment that 'is not comfortable, and it is not easy'. Cross quotes structural engineer for the Sydney Opera House, Ted Happold: 'I really have, perhaps, one real talent; which is that I don't mind at all living in the area of total uncertainty.' It seems there is a creative advantage in risky design practice.

ANOTHER WAY TO DESIGN

Proponents of sustainability often encourage the use of less – do more with less – and that to do so is an indicator of good design. However, this concept when extended further raises the question: Why design anything at all if all it is doing is contributing more stuff to the environment that we don't need? Within the fashion world the idea of less is for most a little frightening. With capitalism itself being based on the ideals of growth – the need for more, faster and cheaper – I don't wonder at the struggle many of us have with having and using less. Despite personally rarely buying new clothes because of the overwhelming guilt I associate with it, I am part of the industry that generates new makers of more stuff. I am cognizant of the fact that if everyone consumed clothing the way I do, most of the students I help educate would not get a job, I would possibly not have a job, and the industry responsible for many things I truly admire would not be in existence. So to ease this personal sense of wrong doing while still enabling the creativity and excitement I love about fashion, I needed to find a solution that aligned my actions with my beliefs. In McDonough and Braungart (2002) I found the beginnings of a way through. They argue that if we change our systems and products then we don't need to make do with less in order to sustain our lives on this planet. The notion that we may not have to do with less to be sustainable certainly appeals to me – that by redesigning the things we buy and use every day we can still enjoy products and services that help enrich our lives. A key concept of cradle-to-cradle design is that we need to eliminate the concept of waste. Not merely reduce waste or down-cycle waste, but to design waste out of the equation and to ensure all components of any product fall into a technical or biological nutrient cycle.

Attempting to design fashion clothing for a technical and biological nutrient cycle led me back to my Master's project, First Son (McQuillan, 2005). In this I designed a series of dresses made from whole pieces of cloth that could be transformed into garments and back again. While my motivation for doing this was not sustainable design, it showed me the potential of designing garments without waste. It also revealed the diverse designs possible within one cut of cloth as each garment could be worn a number of ways depending on how the fastenings were configured. This approach reduces the number of garments required in a wardrobe while satisfying personal desire for variety.

First Son was my first foray into risky design practice. The design evolution of each piece wove like a narrative, starting with one set size and piece of cloth that I cautiously cut into to see how each cut and tuck would influence the next, to craft the design of the dress. Storytelling and memory guided this design process. Sometimes I cut too far and had to start again; it taught me about what actions are reversible on cloth and which are not. Most importantly it taught me that mistakes can be good.

ZERO-WASTE FASHION DESIGN PRACTICE

Zero-waste fashion design is design practice that embraces uncertainty as a way of responding sensitively to both materials and the instability of the environment. It is a step away from egocentric, hierarchical design models that prevail and a step toward a new model for garment design and production, which aims to eliminate the production of waste from the production of clothing.

Zero-waste garment design begins by treating the raw materials of garment production with integrity. A key component of my design practice and research has been actively to pursue the development of an accidental or intuitive design generation process. The fashion industry traditionally aims to minimize risk and play it safe by following predicted social, aesthetic, manufacturing and economic trends and norms in order to be financially successful. This leads to the paradox experienced in fashion where despite seemingly endless choice, most people struggle to find garments that satisfy them for long, resulting in vast volumes of garment and textile waste at both pre-consumer and post-consumer stage. I am constantly exploring my response to these issues through my own design practice, and attempting to find alternative ways of designing and producing fashion. These alternatives are primarily grounded in risky design processes, utilizing variables and limitations not usually employed in the design of fashion. This is in order to subvert the overwhelming desire to create a sort of 'relative cool' – where the value of a fashion item is relative only to its context. By circumventing the usual processes for fashion design and riding high on accidents and serendipitous form not driven by fashion trends, I hope to avoid fads and achieve a representation of what 'rightness' and 'timeless' in design can look and feel like. It reveals what becomes invisible in the consumption of fashion garments – the textile, the hand of the designer, the waste generated and the discarded unfashionable clothes we throw out every new season. It is an absence of absence – nothing is missing. It is a process intent on creating wearable and desirable garments through an integrated design/production/consumption process that uses zero-waste and cradle-to-cradle philosophies.

DESIGN PRACTICE: TESSELLATION

This process is designed with the objective of using one pattern comprised of a tessellating repeat, cut once through multiple layers of different cloths to produce an almost infinite number of possible garment designs. These can be returned to the designer and remade into new garments when the owner becomes dissatisfied or when fashion changes. This system transforms the garment pieces into a form of technical nutrient. A similar modular approach has been used by Arial Bishop (Fletcher, 2008) and Fortune Cookies (Fuad-Luke, 2004, p124), but fit is compromised by the use of straight edges on the modular components and it is not clear if waste is generated during manufacturing, or even if this was a design consideration. Bishop uses an interlocking fastening system cut from the felt itself that appears fragile but does not negatively

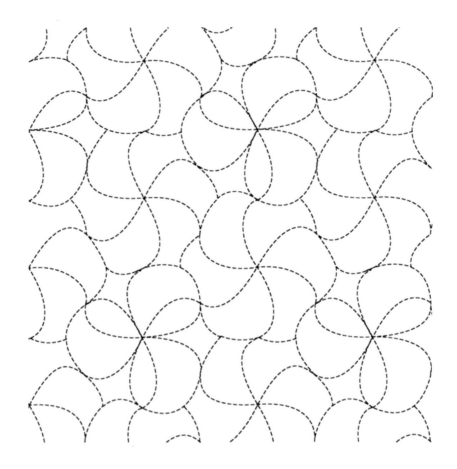

SELVEDGE

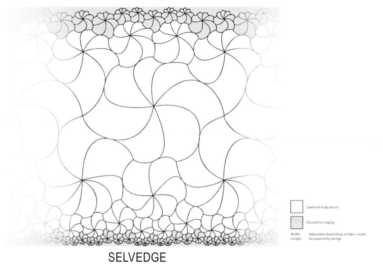

SELVEDGE

LEFT TOP | Figure 2.4.1 Initial tessellation idea

LEFT BOTTOM | Figure 2.4.2a Reducing tessellation design

effect the movement or handle of the fabric, while Fortune Cookies use of Velcro to attach each module to the other renders the fabric stiff and potentially uncomfortable. The fastening system remains the most problematic aspect of this process as constant sewing and unpicking would not only be time-consuming but would degrade the fabric far more rapidly than is usual. So long as a fastening system can be resolved for my work the pieces can be reconfigured a number of times into different garment designs before the material is degraded. This system enables the production of both tailored and fluid designs, depending on the configuration of the pieces and on the fabric used.

Figure 2.4.1 shows my first attempt at using tessellations to generate garment pieces with a delicate balance of form and versatility, while attempting to utilize the whole piece of fabric. The repeat could be scaled up and down to generate a range of aesthetic directions for garments and the design could be entirely different depending on how each piece was arranged and the material used. The shapes that make up the tessellation were designed to respond sensitively to the shape of the body – to curve under arms, wrap around the neck etc. – while still being able to fit into a tessellation. However, as you can see there would be edges that are not incorporated into the tessellation and would need to either be waste or potentially awkwardly incorporated into the garment. I wanted a solution that was by design more intentional.

A solution I found was in mathematics and in particular fractals and hyperbolic tessellations. A commonly known example of the use of mathematics and fractals in art are the works of M. C. Escher, in particular his work with what he called the Circle Limit and what is commonly known as his Reducing Lizards tessellation. The use of fractals borrows from natural patterns and form but its use to me is of a more practical nature. By decreasing the size of the tessellated pieces at the sides there is less fabric waste, while simultaneously giving more variety and options for final garment design. Using nature as a model for design seemed to me to be a logical step in a sustainable direction and represents a sort of economy of evolution that has intrinsic beauty and is naturally environmentally sympathetic.

With the goal of combining the aesthetics and overall form from Figure 2.4.1 within the hyperbolic half plane tessellation format, I generated a repeated pattern that diminishes at the selvedge to generate minimal or no waste, while still creating shapes sympathetic to the curves of the body. This design process is both risky and certain, as I can't predict how the garment will look before I cut the cloth, but as a designer I have control over how to use each piece to make the final design. The process of applying these shapes to a dress form leads to a garment design process more akin to sculpture than drape. Importantly it is an approach that requires a complete shift in the industry towards localized, smaller and slower design approaches and encourages an intimacy between designer, producer and consumer that reunites all three after decades of separation. As a basis for garment design this is a radical shift from existing models of design for clothing, but uses available technology and materials and can be disseminated freely, leading to global design distribution through local production: all this while allowing for aesthetic change with a reduction in material use.

DESIGN PRACTICE: JIGSAW

The Jigsaw design approach is discussed by Rissanen (Hethorn and Ulasewicz, 2008; Rissanen, 2005) and describes a pattern-cutting technique in which all pieces interlock with each other generating no waste from their production. The approach I take develops ideas from the pattern-cutting techniques of Julian Roberts from his formerly online and now book, *School of Subtraction Cutting* (Roberts, 2009) and combines these with the zero-waste design process I used in First Son to generate a range of garments that provides unexpected fluid and organic

89

OVER | Figure 2.4.2b One possible garment design from tessellation

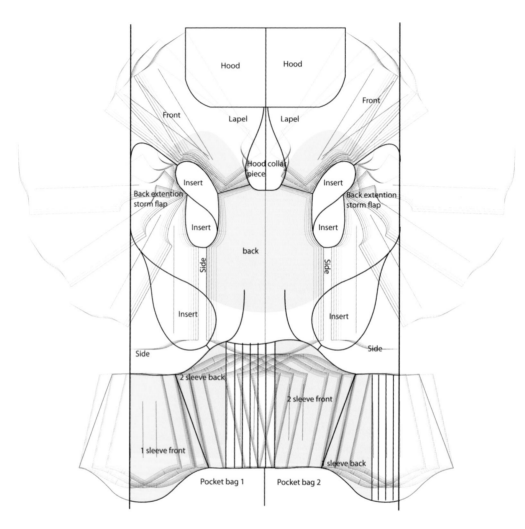

Hood

Hood

Front

Front

Lapel

Lapel

Hood collar piece

Insert

Insert

Back extention storm flap

Back extention storm flap

Insert

Insert

back

Side

Side

Insert

Insert

Side

Side

2 sleeve back

2 sleeve front

1 sleeve front

1 sleeve back

Pocket bag 1

Pocket bag 2

ABOVE | Figure 2.4.3 Design in progress, showing established fixed areas in yellow

forms. Utilizing Roberts' 'plug' technique – whereby any shape can be inserted into any void so long as the diameter of both are the same – enables any part of the garment that is removed for fit or aesthetics to be reincorporated into the design of the garment. This creates unexpected outcomes that encourage both the wearer and designer to take risks. Through this design process the garment and pattern (and occasionally the textile print) are designed simultaneously, dealing with two-dimensional pattern and three-dimensional garment form at a single production stage. Such an approach requires that the designer be both a proficient pattern cutter and designer in order to be successful.

Rissanen (2007) discusses three designers, Julian Roberts, Yeoh Lee Teng and Yoshiki Hishinuma, who design following a similar process whereby the pattern (through a process of refinement and toiling) is the originator of the garment design. He writes, 'it seems that the three designers, while allowing a degree of uncertainty into their designing, decide on the type of garment (dress, jacket, trousers) before the pattern is made'. I begin my design process with two additional guides, the first being the width of the fabric I am using for the particular design I am working on, and the second being a 'fixed' area to begin the design process from. A fixed area might be a decision on the design of a sleeve, or how the neckline might be shaped or how the garment might fit the waist or shoulder area. This fixed area is where all other design decisions radiate out from, and aside from the general garment type is the only predetermined aspect of any given design. This fixed concept can be applied within a single garment a number of times, but the more there are and the larger the fixed areas are, the more difficult the design process will be and it will result in a less spontaneous design. Aside from these two boundaries all other aspects of the design are fluid at this point. The goal is not to minimize the fabric used, but to eliminate the concept of waste in garment creation. This is an intentional creative decision, as while I don't aim to be ostentatious with my fabric use, I will not further restrict myself to a set length of fabric when I begin to design. This distinction is a concept that McDonough and Braungart identified as a key direction in their cradle-to-cradle philosophy. Every designer at some point will reach a limit in terms of restrictions they choose to adhere to. No waste happens to be mine.

Aside from the fixed area, I cannot know how the end result will appear; although the more familiar the designer becomes with this process the easier it is to envisage the result as the design develops. By remembering that any shape will sew into any void as long as the circumference is compatible, the design process begins to form around these fixed areas. It is important to be mindful that with every line you draw when developing a zero-waste pattern you are in fact designing two pattern pieces – and when you cut the fabric you are cutting both sides of the scissors. Arbitrary decisions therefore are not possible as every action has an equal and opposite reaction – a reality that intentionally initially slows the design process down. Designing garments and patterns in this way requires a shift in focus – from knowing what you want the end result to be and exactly how you will achieve it, to being courageous, conscious and thoughtful as you design the pattern or garment in order to meet generalized and flexible goals of fit, aesthetics and importantly waste elimination.

93

Australian designer Mark Lui employs a textile laser cutting technique to use the space between more standard garment patterns, so that the laser cut seams become exterior decoration and detail. The addition of innovative textile design makes the possibilities for zero-waste garment production almost endless. Both laser cutting and digital textile printing lend themselves well to this process and can become new tools for garment design. Aldrich (1996, p5) wrote that 'the tension between precision and the speculative cut of new fabrics can generate new garment forms': this statement supports the approach where restriction and freedom precariously balance against one another to lead to innovative garment aesthetics.

DESIGN PRACTICE: EMBEDDED JIGSAW

One criticism of the Jigsaw method is that it can lead to designs that have an abundance of fabric and drape – which is not always desirable. One way around this is to 'embed' a traditionally designed garment pattern into a zero-waste pattern and treat the embedded pattern as a 'fixed' area. This enables multiple garment designs and types from a single zero-waste pattern. Figure 2.4.4 shows a menswear hooded jersey and T-shirt within the same pattern, using a digital textile print to differentiate between the two garments. The design process is the same for embedded designs as for traditional jigsaw approaches. MaterialbyProduct used a similar method when they placed a simple traditional dress pattern from a length of cloth and the negative space draped to generate a more complex design, resulting in 100 per cent yield over the two dresses. This process enables more fitted or defined individual garment forms while still achieving zerowaste.

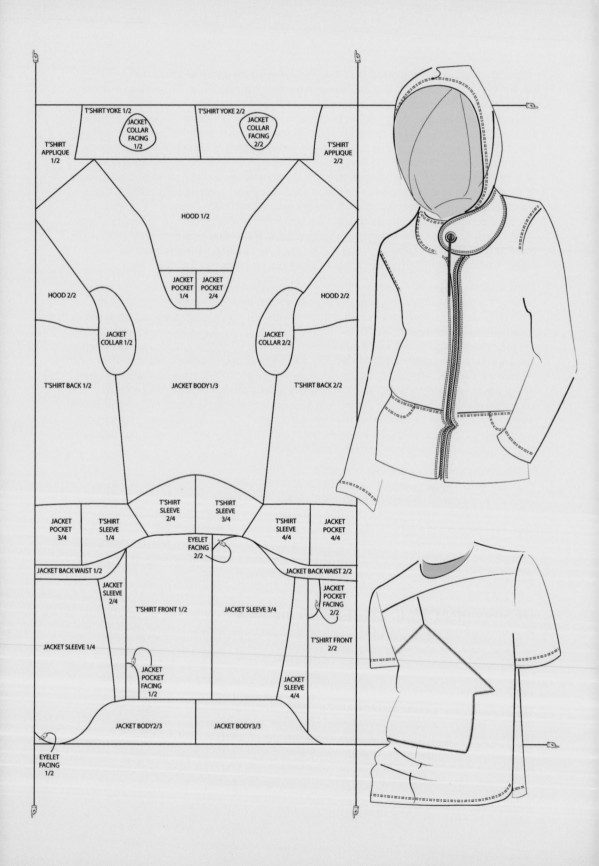

T'SHIRT YOKE 1/2

T'SHIRT YOKE 2/2

JACKET COLLAR FACING 1/2

JACKET COLLAR FACING 2/2

T'SHIRT APPLIQUE 1/2

T'SHIRT APPLIQUE 2/2

HOOD 1/2

HOOD 2/2

HOOD 2/2

JACKET POCKET 1/4

JACKET POCKET 2/4

JACKET COLLAR 1/2

JACKET COLLAR 2/2

T'SHIRT BACK 1/2

JACKET BODY1/3

T'SHIRT BACK 2/2

T'SHIRT SLEEVE 2/4

T'SHIRT SLEEVE 3/4

T'SHIRT SLEEVE 4/4

JACKET POCKET 3/4

T'SHIRT SLEEVE 1/4

EYELET FACING 2/2

JACKET POCKET 4/4

JACKET BACK WAIST 1/2

JACKET BACK WAIST 2/2

JACKET SLEEVE 2/4

JACKET POCKET FACING 2/2

T'SHIRT FRONT 1/2

JACKET SLEEVE 3/4

JACKET SLEEVE 1/4

T'SHIRT FRONT 2/2

JACKET POCKET FACING 1/2

JACKET SLEEVE 4/4

JACKET BODY2/3

JACKET BODY3/3

EYELET FACING 1/2

LEFT AND ABOVE | Figure 2.4.4 Hoody/T-shirt embedded design; pattern, line drawing, rendering of possible digital print colourways

DESIGN PRACTICE: MULTIPLE CLOTH APPROACH

A similar approach that enables more variety and flexibility within zero-waste garment cutting, is to design two or more patterns for different fabrics at the same time. This enables the designer to embed cloth from one pattern into the design of another in multiple possible configurations. The same approach could be used for a single pattern cut from two or more different lengths of cloth, to develop a variety of colour and texture without needing to design different patterns.

All of these zero-waste methods are designed to encourage trial and error and risk taking in the design process. Combining different techniques from within traditional pattern-cutting, drape and zero-waste garment design processes open up more opportunities for development. In addition, combining different zero-waste processes, such as Jigsaw with tessellation, could develop down lines similar to the embellishment approach designers such as Mark Lui and Alabama Chanin use. The garment design processes articulated in this paper add to the practices emerging from the zero-waste design practice movement, showing that the only restriction to what is possible is your imagination.

MAKING

In all of these processes the hand of the designer is extended to pattern-making and into construction. Trailblazing designer Julian Roberts writes of his relationship to sewing:

> When I make garments I manipulate the cloth by hand as I join together seams, working with the fabric's natural drape and easing seamlines evenly so they fit. I sew in bursts and rhythms, using the under-teeth and knee pedal to walk/run the fabric through the machine, fingering the fabric carefully beneath the needles tip. I understand the tolerance of each fabric, the subtleties of easing and stretching, pivoting and topstitching... My hands and fingers are all over my women's garments. I touch every inch of them, my DNA is everywhere, sewing isn't something I prefer to delegate, it is part of the cutting, part of the design, and if I do work with other machinists, I need them to understand and be able to demonstrate to them the high level of finish necessary, and how much of their hand skills will be treasured in each garment. (Roberts, 2010).

95

Julian Roberts' description of sewing the garments he designs, reminds us of all the hands involved in making both the clothes we wear and the ones we design. Just as divorcing consuming from producing has led to overconsumption, the divorcing of design from production has led to decisions being made without forethought as to their implications. When considering that 80 per cent of a product's environmental impact is determined at the design stage (Thackara, 2005) it becomes clear that designers need to re-engage with the whole fashion system, at all levels and in all ways to make real change in the industry.

CONCLUSION

Fast fashion is a system that taps into our desire for change and fulfils it at a price point we can easily justify spending our hard earned or borrowed money on. We need to design new fashion systems, which fulfil this desire for flexibility and change but which do not impact on our environment and society so severely. Therefore the future of the fashion industry cannot lie in organic garments within the traditional fashion system, whose production still generates hundreds of thousands of tonnes of textile waste, garments that then get transported around the world thousands of kilometres only to be discarded into landfill at the end of the season. While organic fibres are a part of the solution, we need to re-make the way we make and consume clothes. So how is this possible? It is in the development of innovative clothing design, production and consumption systems that can be disseminated freely and adapted for anywhere in the fashion world. A new fashion system that builds on the developments already made in the areas of organic and fair-trade industries and works to link producers with consumers and fosters original design and flexible aesthetics. I have focused on exploring new ways of clothing creation that could serve as a transition process between old and new. The concept of zero-waste pattern-cutting is in many ways a relatively new idea. In the past when cutting a garment such as the Kimono, aiming for little or no waste would be assumed and did not need to be defined or labelled as such, it was just how things were done. Zero-waste design practice can help to slow down the design process and begin to cause some change toward slowing the fashion system as a whole. Additionally it can also be utilized in an otherwise sustainable fashion company to further reduce waste and resource misuse, while generating truly innovative forms. It is where well-being meets the well-dressed through the resolution of the conflict inherent in the notion of sustainability and the current model of fashion consumption, by embracing unpredictable design processes as a creative advantage – resulting in confronting new ways of designing, producing and owning clothes.

REFERENCES

Abernathy, F. H. (1999) *A Stitch in Time: Lean Retailing and the Transformation of Manufacturing – Lessons from the Apparel and Textile Industries*. Oxford University Press, New York

Aldrich, W. (1996) *Fabric, Form and Flat Pattern Cutting*, John Wiley, Chichester

Anon. (2008) Personal correspondence to the author, 10 December

Boza, L. E. (2006) '(Un)intended discoveries', *Journal of Architectural Education*, vol 60, no 2, pp4–7

Burnham, D.K. (1973) *Cut my cote*. Royal Ontario Museum, Canada

Chapman, J. and Gant, N. (eds) (2007) *Designers, Visionaries and Other Stories: A Collection of Sustainable Design Essays*. Earthscan, London

Cooklin, G. (1997) *Garment Technology for Fashion Designers*. (Illustrated edition) Wiley-Blackwell, Oxford

Cross, N. (1999) 'Natural intelligence in design', *Design Studies*, vol 20, no 1, pp25–39

Feyerabend, R. (2004) 'Textiles briefing paper'. www.mrs-hampshire.org.uk/Workshop%204/Textiles.pdf (accessed 18 June 2010)

Fletcher, K. (2008) *Sustainable Fashion and Textiles*. Earthscan, London

Fuad-Luke, A. (2004) *The Eco-Design Handbook: A Complete Sourcebook for the Home and Office* (2nd edition). Thames and Hudson, London

Hethorn, J. and Ulasewicz, C. (2008) *Sustainable Fashion: Why Now?: A Conversation Exploring Issues, Practices, and Possibilities*. Fairchild Publishers, www.nordes.org/data/uploads/papers/122.pdf (accessed 27th August 2010)

McDonough, W. and Braungart, M. (2002) *Cradle to Cradle: Remaking the Way We Make Things*. North Point Press, New York

McQuillan, H. (2005) First Son: Memory and myth – an adjustment of faith: A written component presented in partial fulfilment of the requirements for the degree of Masters of Design in Fashion and Textile Design, Massey University, Wellington, New Zealand

Rissanen, T. (2005) 'From 15% to 0: Investigating the creation of fashion without the creation of fabric waste'. www. kridt.dk/conference/Speakers/Timo_Rissanen.pdf (accessed 5 February 2007)

Rissanen, T. (2007) *Types of Fashion Design and Patternmaking Practice*. Design Inquiries/Nordes Stokholm, www. nordes.org/data/uploads/papers/122.pdf (accessed 27th August 2010)

Roberts, J. (2009) *School of Subtraction Cutting*. Center for Pattern Design. www.centerforpatterndesign.com/products/SCHOOL-OF-SUBTRACTION-CUTTING-by-Jullian-Roberts.html (accessed 12 May 2010)

Roberts, J. (2010) *JULIANAND Productions*. www.julianand.com/WhatDoYouWant.htm (accessed 12 May 2010)

Santi, A. (2008) 'Throwaway fashion grows to 30% of landfill waste'. www.drapersonline.com/news/throwaway-fashion-grows-to-30-of-landfill-waste/1933004.article (accessed 11 October 2009)

Thackara, J. (2005) *In the Bubble: Designing in a Complex World*. The MIT Press, Cambridge, MA

Tiplady, R. (2006) 'Zara: Taking the lead in fast-fashion', BusinessWeek: Europe. www.businessweek.com/globalbiz/content/apr2006/ gb20060404_167078.htm?chan=innovation_branding_brand+profiles (accessed 15 June 2010)

Tokatli, N. (2007) 'Global sourcing: Insights from the global clothing industry: The case of Zara, a fast fashion retailer', *Journal of Economic Geography*, vol 8, no 1, pp21–38

USE
Chapter 3

USE | INTRODUCTION

This chapter centres on the use phase of clothes: wearing them, laundering and drying them, and maintaining them. Often fashion's most significant environmental impacts are created during the use phase; primarily from too frequent laundering, laundering in unnecessarily warm water, laundering small loads and using tumble driers to dry clothes. As the fashion designer is not in direct contact with the consumer during this phase, addressing these issues through design may seem difficult. This chapter aims to demonstrate that fashion design can indeed impact upon consumer behaviour towards a more environmentally considered use phase, and perhaps more engaged relationships with fashion amongst consumers.

Fletcher (2008) notes that technological innovations, such as more efficient washing machines, cannot alone solve the impacts arising from the use phase. Laundering is strongly associated with cultural and social constructions of cleanliness and convenience. With advances in laundering and drying technologies over the past 60 years, the perceived acceptable levels of hygiene and convenience have risen, resulting in significantly increased impacts associated with laundering and drying clothes (Shove, 2003). Kathleen Dombek-Keith and Suzanne Loker place the environmental impacts of laundering in a design and manufacture context, and identify strategies that designers and manufacturers may be able to adopt to educate consumers and guide them towards more sustainable behaviours in regards to laundry. The authors make a strong case for garment care as an integral aspect of fashion design, thus expanding the conventional notion of what fashion design entails.

This chapter examines the use phase also in terms of the kinds of relationships consumers have with clothes. As the prices of clothing have fallen, these relationships have tended to become increasingly fleeting. As a product category clothing is one that the consumer has one of the most intimate relationships with, as clothing acts as a protective layer between the wearer and the world. Clothes eventually wear out or more commonly the consumer may grow bored with a garment, leading to a somewhat prematurely shortened use phase and higher levels of textile waste. Focusing on visual and physical durability, Timo Rissanen argues that fashion design and pattern-making can facilitate transformative practices such as repair and alteration during the use phase, potentially extending the use life of a garment. Visibly repaired or altered clothing is still often perceived as possessing less value, although at times this may emerge as a fashion trend, particularly in denim garments. Rissanen proposes that designing clothes with an element of disturbance may make a future transformation of a garment, such as mending or altering it, more acceptable for the wearer. Furthermore, incorporating fabric that is ordinarily wasted into a garment can result in increased physical durability.

....

Fletcher, K. (2008) *Sustainable Fashion & Textiles: Design Journeys.* Earthscan, London

Shove, E. (2003) *Comfort, Cleanliness + Convenience. The Social Organization of Normality.* Berg, Oxford and New York

KATHLEEN DOMBEK-KEITH AND SUZANNE LOKER

3.1 SUSTAINABLE CLOTHING CARE BY DESIGN

INTRODUCTION

Designers have often aspired to radicalism... Only by re-thinking some basic assumptions about function, tastes and lifestyle will we be able to move any significant way towards a more sustainable way of living (Mackenzie, 1997, p168).

A primary goal for any eco-friendly design initiative is developing a system of sustainable consumption, that is, the production and use of goods and services. In 1992, the United Nations' Agenda 21 (UN, 2004) first considered sustainable consumption when it called for 'new concepts of wealth and prosperity, which allow higher standards of living through changed lifestyles (Chapter 4.11). In order to accomplish the goals of Agenda 21, Jackson and Michaelis (2003) argue that consumers must change their lifestyle expectations and behaviours. Unfortunately, many current sustainable design initiatives focus on the production side of sustainable consumption systems rather than on product use and care that are ultimately the consumer's responsibilities. Designers can serve as the bridge between production and use, addressing the needs and concerns of manufacturers and consumers in the design stage. The choices made in the design stage can be either sources of effective sustainable strategies or causes of greater negative environmental impact when designers fail to consider the effects of existing and potentially new consumer use behaviours (Lilley et al, 2005). Therefore, eco-friendly fashion designers must understand consumers' 'normal' actions and standards of acceptable clothing care and use in order to discover new, effective ways of reducing resources used in overall clothing consumption, such as promoting low-impact care and enabling extended clothing use, resulting in sustainable clothing care by design.

A SYSTEMS ANALYSIS OF CLOTHING CARE

Shove (2003, 2004) contends that to truly understand how consumers make choices about laundering, a system of individual systems must be evaluated (Figure 3.1.1), including: what are the tools for laundering, what is laundered, and why, when, how, and by whom laundry is done. Answers to these questions can provide the insights necessary to bring about changes in how designers design and consumers consume that can reduce the environmental impact of clothing care.

A 'shared understanding of "normality"' as it relates to laundering practices and standards of cleanliness can reveal the underlying benefits that consumers think laundering provides and influence their care choices and behaviours (Shove, 2004, p77). For example, switching to line drying would significantly lower laundering energy costs, but since most households in the US machine dry, it must offer certain perceived consumer benefits not provided by line drying. In order for consumers to accept line drying, they must find methods of line drying that mimic the benefits of machine drying and/or alter their acceptance of benefits particular to line drying. Laundering goals and preferences have changed dramatically over time, reflecting the evolution of everyday life. A few centuries ago when bathing was rare, changing clothing soaked with bodily odours and stains substituted for washing oneself. Now in the cleanliness of modern life, laundering is viewed as a process of whitening and freshening clothing from soil and odours to make them acceptable for touching our clean bodies (Shove, 2003, 2004). This ever-evolving definition of laundering does offer hope that consumers' current conceptions of 'normal' laundering behaviours are open to change with the help of the designer's influence.

Acceptable clothing care reasons and methods are social constructions based on the habits of friends and families as well as product availability and advertising that have focused at different times on garment disinfection, whiteness, softness or other desirable characteristics. Reasons behind normally acceptable laundering frequency can depend on the collective understanding within a social group. For example, it may be after one day of wear, for reasons of odour rather than clean appearance. To address this social norm, the designer may consider methods of eliminating offensive odours that focus on freshening and use fewer resources than washing. Considering the integrated systems of laundering together (i.e. tools, why and when) may lead designers to novel design strategies that change consumer behaviour options and selections toward more sustainable clothing care. Using Shove's systems approach as a foundation, we will explore current consumer clothing care options, including equipment and detergent choices, general care practices and socially defined purposes for laundering, to inform the designer's role in evaluating and addressing the environmental impacts of the clothing care system. Then, we will present design approaches for promoting and enabling sustainable clothing care and use lifestyles that designers can use to change or incorporate into the fashion system in order to reduce resources used in overall clothing consumption.

ASSESSING CURRENT CLOTHING CARE LIFESTYLES

The 'use' stage of a garment's life typically demands the most environmental resources, up to 80 per cent of an everyday garment's lifetime energy use (Collins and Aumônier, 2002). Lifestyle factors that contribute to the amount of energy, water and other resources used for clothing care include laundering frequency and load size, water amount and temperature for washing, laundry detergents and additives for cleaning and improving clothing appearance, heat used in tumble drying and garments that require special care. Fortunately, it is within the consumer's power to substantially reduce these impacts through basic behavioural changes and investments in energy-efficient laundry machines, thereby lowering household energy bills and overall resources used during a garment's lifetime.

Laundering Frequency: Reducing Consumers' Need to Clean

A simple yet substantial way to reduce the resources used for clothing care is to do fewer loads of laundry. Accomplishing this goal requires understanding consumers' reasons for doing

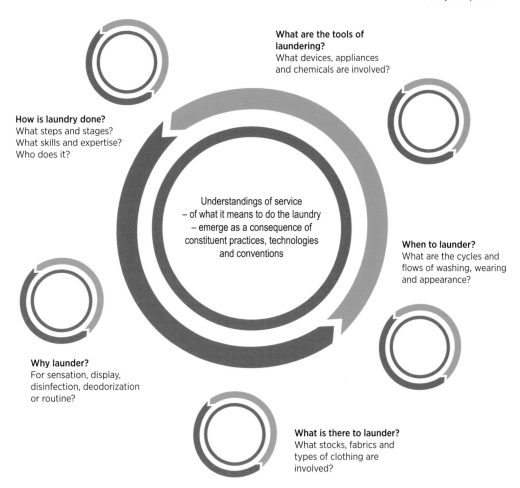

What are the tools of laundering?
What devices, appliances and chemicals are involved?

How is laundry done?
What steps and stages?
What skills and expertise?
Who does it?

Understandings of service
– of what it means to do the laundry
– emerge as a consequence of
constituent practices, technologies
and conventions

When to launder?
What are the cycles and flows of washing, wearing and appearance?

Why launder?
For sensation, display, disinfection, deodorization or routine?

What is there to launder?
What stocks, fabrics and types of clothing are involved?

ABOVE | Figure 3.1.1 Laundry as a System of Systems. Shove (2003)

laundry and perhaps altering them. For example, different social definitions of 'dirty' can render a garment unfit for further wear either after minimal use or not until it has noticeable stains or odours. Over the past decades, load frequency has increased, perhaps due to wardrobe size growth, the convenience of laundering with household machines and consumers' notions of cleanliness. As of 2005, US households with washing machines washed an average 315 loads of laundry per year; that is 6 loads per household or over 2 loads per person each week, based on our calculations of survey data (US DOE, 2005). Reducing the number of loads washed by 50 per cent would halve use of energy, water and laundry additives and could reduce energy consumption over a garment's lifetime by 15–30 per cent (Allwood et al, 2006). Combining several smaller loads into fewer full loads would decrease laundering frequency and save consumers time and money. For example, increasing a load size from 3kg to 3.5kg can reduce laundry energy use by 14 per cent and an individual garment's lifetime energy use by 5 per cent (Collins and Aumônier, 2002). Consumers could also extend wear time between washings by loosening their definitions of what is and is not dirty, using spot removers like 'Tide to Go' to remove stains without washing, hanging unstained worn clothing to air out rather than throwing it in piles, or purchasing garments with anti-stain or anti-odour finishes that do not easily get dirty or smelly.

Washing Machines: Saving Water and Energy

The choices consumers make when selecting their washing equipment and wash cycle options can greatly affect the amount of water and energy used per load. The majority of energy needed to wash a load of laundry is for heating the water, so energy use increases as the water's quantity and temperature rise. Standard top-loading washing machines require up to 151 litres of water to completely cover the clothes in the basin, so heating that amount of water can require up to 90 per cent of the energy used per load. Newer high-efficiency top- and front-loading models that meet qualifications for Energy Star labelling use only 68–95 litres of water and up to two-thirds less energy per load. Front-loading models typically offer the most water and energy savings as they only need to fill the bottom of the washer basket as the clothes rotate horizontally in and out of the water, and they are gentler on clothes than standard top-loading washers with central agitators that snag clothing. High-efficiency top-loading machines reduce water use through modified spray cycles and smaller basins with smaller or no agitators. Both top- and front-loading high-efficiency washing machines are designed to spin much faster than standard top-loading models, so more water is removed from the clothes resulting in 25–30 per cent faster drying times (Flex Your Power, 2010b; Ashley, 1998).

In 2005, only 24 per cent of US households with washing machines owned Energy Star models, while 60 per cent owned washers older than four years and 11 per cent owned newer models that were not Energy Star rated (US DOE, 2005). Yet some basic consumer practices can reduce resource use with any washer, such as matching the water level to the load's size or just washing full loads. Changing the washing water temperature can also result in considerable energy savings: switching from Hot (50°C/122°F) to Warm (40°C/104°F) can decrease overall energy use during a garment's lifetime up to 10 per cent (Collins and Aumônier, 2002). Rinsing in cold water is sufficient to wash away soapy water. US DOE survey data (2005) show that households with washing machines most often wash in Warm (55 per cent) followed by Cold (34 per cent), and 79 per cent rinse in Cold, so many US consumers already follow some low-impact washing practices. In the future, washing machines may use cleaning solvents other than water and operate to conserve resources. Sanyo now offers an innovative (and expensive) washing machine that uses ozone made from oxygen in the air to clean clothing by oxidizing stains and odours, and water to rinse that is recycled after cleansing with ozone (Dunn, 2006).

Laundry Detergents and Softeners: Effectiveness and Environmental Concerns

Laundry detergents and softeners provide consumers with desired benefits of contaminant removal, pleasing 'fresh' scents, whiteness and fabric softness, yet their ingredients demand resources for functionality and pose environmental concerns such as toxicity and pollution. Many laundry detergents include the same two key ingredients: surfactants and builders. Surfactants, or surface-active agents, are the primary active cleaning ingredients that bind to water and soil thereby extracting and rinsing it away from the clothing (Cameron, 2007). Some surfactants degrade into toxic chemicals that persist in the environment, but eco-friendly varieties biodegrade into safe by-products (US EPA, 2010). Surfactant functionality depends on its molecular charge (negative, neutral and positive) and water quality. Hard water contains positively charged contaminants that can disrupt the chemical interactions of surfactants, especially those that are negatively charged. For this reason, base solution builders are often added to detergent formulations to neutralize acidic hard water. In addition, builders improve the detergent's cleaning performance by loosening and suspending soil from the clothing, reducing the amount of surfactants needed resulting in lower product costs (Frydendall, 2010). Phosphates were highly effective builders widely used in detergents before the 1990s but were found to cause eutrophication, or increased nutrients, when entering aquatic ecosystems through wastewater. The resulting 'blooms' in algae populations would deplete the oxygen from water, making it uninhabitable for aquatic life. Consequently, most US manufacturers replaced inorganic phosphates with organic zeolites that are considered safer for the environment while still effective at treating water (US EPA, 2010).

Concentrated and cold-water detergent formulations were developed to conserve resources. Concentrated detergents contain fewer fillers, typically sodium sulfate in powders and water in liquid detergents (Frydendall, 2010). Resources are saved through smaller package sizes and weights, reducing transportation costs for the same amount of wash loads. Effective cold-water detergents were difficult to achieve, requiring balancing biodegradability with solubility at low temperatures (McCoy, 2003). They typically have formulas with more hydrophobic surfactants that better attach to grease and oil and include 20 per cent more active ingredients for the same cleaning power as standard detergents (Petkewich, 2005). Washing in higher water temperatures is still more effective for breaking up tough stains and disinfecting garments, but lightly soiled clothing can be effectively cleaned in cold water with the appropriate detergents.

Common detergent additives that provide desirable laundry qualities include bleaches, whiteners and fragrances. Bleaches are used for stain removal, disinfection and often whitening. Chlorine in bleaches can degrade into persistent toxic by-products and damage fabric fibres and colours, yet non-chlorine bleaches, like hydrogen peroxide and ozone, are non-toxic and safer for clothing and the environment. Whiteners or optical brighteners are designed to make whites look brighter, often by adding a tint of blue to yellowed clothes. However, most whiteners can be harmful to human health and leave toxic, persistent by-products in the environment. Fragrances are added to detergents either to mask the chemical smells of other ingredients or to provide a pleasing 'fresh' scent. Fragrances can be volatile organic chemicals (VOCs) that pose health concerns, including headaches, breathing problems and skin irritations when used in high doses or for individuals with sensitivities, as well as environmental concerns about potentially polluting by-products (US EPA, 2010; Steinemann, 2009).

105

Fabric softeners offer many benefits consumers desire by lubricating laundry to improve fabric feel or 'hand', reducing static electricity during tumble drying, and making it easier to remove wrinkles during ironing. Yet the need for fabric softeners has been an unintentional consequence of washing machines and laundry detergents that better remove oils and other lubricants from fabrics leaving them feeling harsh after washing, especially those made with naturally oily fibres like cotton. Most fabric softeners are made with natural oils from tallow, or animal fats, and plant seeds, so they are non-toxic and biodegradable. However, fabric softeners often have additional ingredients that can be toxic and polluting, like fragrances. Softeners can be added to laundry during washing as detergent additives or separate solutions, or during tumble drying as dryer sheets that remove static cling. Fabric softeners in either form help reduce drying time, which can translate into lower energy costs and longer clothing lives (Levinson, 1999).

Machine vs Line Drying: Energy Costs vs Consumer Needs

How consumers choose to dry their laundry is a calculation of costs, comfort and convenience, and for US households, this often results in favour of machine drying. Yet more consumers are considering the economic and environmental costs of inefficient tumble dryers, especially when compared to energy-free clotheslines (Morris et al, 1984). Machine drying can account for up to 60 per cent of all energy used during the care stage of a garment's life (Allwood et al, 2006) and around 6 per cent of total US household energy use (Flex Your Power, 2010a). The energy source powering the dryer significantly affects consumer and environmental costs. Most US household clothes dryers are electric, but 17 per cent are powered by natural gas, which is cleaner than coal-based electricity resulting in lower carbon emissions. Additionally, gas dryers can cost 50 per cent less to operate over their lifetimes, despite higher initial appliance costs due to gas hook-up installation (Flex Your Power, 2010; US DOE, 2005). Almost all machine dryers are vented models that blow hot air into the tumbler to evaporate water off wet clothes and then the hot, humid air is released through the vent. Vented dryers must operate at very high temperatures to maintain evaporation throughout the drying cycle, causing garment shrinkage and shortened lifespan, though some models have moisture sensors that shut off the dryer when clothes are dry to prevent over-drying that further damages clothing and wastes energy (Ashley, 1998). Yet this basic design has little opportunity for further energy efficiency so the US Environmental Protection Agency (US EPA) does not require vented dryers to carry energy labels and no Energy Star options exist.

Emerging technologies for machine dryers offer new opportunities for significant energy and time savings, yet considerably higher product costs and challenging technical issues must be overcome before mass-market availability (Flex Your Power, 2010a). Microwave-based dryers promise 17–25 per cent reduction in energy consumption, 25–50 per cent faster drying times and lower operating temperatures, making it safer to dry delicate fabrics like wool (Flex Your Power, 2010a; Gerling, 2003; Ashley, 1993). Microwaves work by heating water directly so it evaporates without having to heat up the clothing (Ashley, 1993). Troublesome issues include electric arcing and fabric scorching from metallic objects that are part of or left in clothing, but various solutions are under development (Gerling, 2003). New ventless dryers utilize condensation to remove water from the exhaust so that once-wasted hot air can be recycled back into the tumbler, potentially reducing overall energy use by 50 per cent. Steam compressor-based dryers heat up quicker than vented dryers so loads dry up to 40 per cent faster (Palandre and Clodic, 2003). Microwave and ventless drying machines also help reduce energy costs by eliminating outdoor vents through which heated or cooled building air escapes.

Complete abandonment of machine drying for line drying could result in the most significant energy savings over the garment's entire life, up to 25 per cent (Collins and Aumônier, 2002). Unfortunately, US consumers are in the habit of machine drying their laundry: in 2005, 83 per cent of US households containing a clothes dryer dried every load of laundry (US DOE, 2005). Realistically, line drying may supplement but not replace outright machine drying due to weather conditions, space and time constraints, and consumer preference for soft, wrinkle-free clothing. Heat from machine drying helps remove wrinkles caused by machine washing and softens fabric hand, especially natural fibres like cotton. Furthermore, clotheslines are scarce in many communities across the US due to regulations banning them for aesthetic reasons, though this is changing (Morris et al, 1984). Project Laundry List (2007) works to eliminate such restrictions and promote clothesline use by presenting the energy cost savings, environmental impact reduction and laundry quality benefits of line drying.

A 1984 study by Morris et al tested consumers' clothing and care preferences to determine ways of promoting line drying that designers could apply. Participants cited three main reasons for not line drying their laundry: the convenience of machine drying (62 per cent), dissatisfaction with the hand and appearance of line-dried clothing (38 per cent) and not having a clothesline due to lack of space or unsightly appearance (21 per cent). Machine-dried and line-dried fabric samples of various cotton and polyester blends and fabrications were compared. None of the line-dried samples was found acceptable by a majority of the participants, yet acceptance of line-dried laundry increased when participants were previously informed about the cost savings and environmental resources conserved as a result of line drying laundry. The study then tested methods for improving the hand and appearance of line-dried clothing and found that machine drying for 15 minutes (5 minutes of heat and 10 minutes cool down) after line drying resulted in acceptable ratings almost as high as machine-dried laundry while saving 75–90 per cent in energy costs (Morris et al, 1984).

Special Care Laundry: Environmental Impacts and Changing Consumer Demands

Since the advent of household laundry machines and wash-and-wear fibres and finishes, most clothing and household textiles are washed and dried by machine (Morris et al, 1984). However, some laundry, typically formal attire and linens made from delicate fibres like wool, silk and rayon, still requires specialized cleaning, gentler handling and extra finishing (i.e. pressing) provided by dry cleaning services. The dry cleaning industry suffers from a poor environmental record due to its use of the cleaning solvent perchloroethylene, or 'perc', a VOC, responsible for the chemical smell on recently dry-cleaned goods (The Mediacenter, 2008; Dos Santos, 2007). Perc is an air pollutant and groundwater contaminant that poses health risks to industry workers and consumers from ailments such as acute skin and respiratory irritations to chronic organ damage and cancer (Dos Santos, 2007). While the vast majority of US drycleaners still use perc, regulations and equipment upgrades that prevent spills and reclaim the solvent have reduced environmental contamination (The Mediacenter, 2008). The US dry cleaning industry also has been affected by workplace trends toward more casual clothing that does not require special care and, recently, by the economic downturn causing many consumers to forgo luxuries like dry cleaning services.

In order to survive, the dry cleaning industry is developing and implementing 'greener' yet effective cleaning methods (The Mediacenter, 2008), although some are more sustainable than others. The 'GreenEarth' method is a silicon-based solvent, yet its manufacturing requires chlorine and it is suspected to carry a cancer risk (Dos Santos, 2007). Some drycleaners use non-toxic and non-flammable liquid CO_2 that is inexpensively recycled from industry by-products as a cleaning solvent, although its adoption is limited due to requiring specialized, expensive machinery. Wet cleaning is a more commonly available eco-friendly cleaning method that is also non-toxic, using water as the cleaning solvent and water-based stain removers. It is more energy efficient than dry cleaning as it does not require toxic solvent reclamation equipment and uses less water overall, although it does produce wastewater. Wet cleaning also requires specialized, computer-controlled washing and drying machinery that provides gentle care through slower spin speeds and customizable heat and time cycles, yet they are affordable and have lower operating costs than dry cleaning machinery. Initially, wet cleaning was not considered to clean as well as dry cleaning, but technological advances have improved cleaning so that most 'dry clean only' garments can be effectively wet cleaned, although, consumers could save money by carefully hand washing delicate garments (Dos Santos, 2007).

SUSTAINABLE CLOTHING CARE BY DESIGN

Designers operate within a fashion system that, along with the clothing care system outlined by Shove (2004), play separate yet interconnected parts within the overall system of clothing consumption. Designers' leading roles within the fashion system can enable them to be change agents addressing the impact of clothing care by informing consumers about issues and helping them make changes to their clothing care and use practices that reduce environmental impact. Designers can achieve these goals by implementing design approaches that promote low-impact laundry practices through innovative materials and garment designs that extend the wear life of a garment through versatility, adaptability and meaningfulness. They can also define sustainable clothing care and low environmental impact as part of their business missions and present it as their design inspiration or vision. Yet the ultimate sustainable approach to clothing design may be to make sustainability a fashion itself – to design clothing that symbolizes sustainable care in ways that are easily recognizable by all.

GREEN CLOTHING CARE

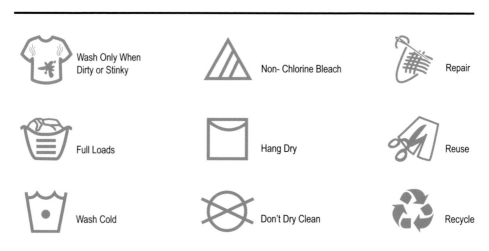

Wash Only When Dirty or Stinky	Non- Chlorine Bleach	Repair
Full Loads	Hang Dry	Reuse
Wash Cold	Don't Dry Clean	Recycle

108

ABOVE | Figure 3.1.2 Green Clothing Care (GCC) Label. Dombek-Keith (2009)

Promoting Consumer Awareness of Clothing Care Impacts: Education, Provocative Designs and Sustainability as Fashion

Consumers who are aware of laundering's heavy environmental impacts and costs are more likely to find sustainable clothing care lifestyles acceptable (Morris et al, 1984). Care labels provide an established medium where designers have the opportunity to directly communicate with consumers. Originally, care labels were developed to benefit clothing manufacturers: first through written care instructions that offered legal protection from consumers who improperly laundered their products, and then through care symbols that were supposed to overcome written language limits to allow for expanded consumer markets (Federal Trade Commission, 1995). Yet many consumers misunderstand care symbols, especially symbols that do not clearly represent their meanings, such as a triangle for bleach (Workman and Choi, 1999). Designers can utilize and expand care labels to benefit consumers and the environment by promoting low-impact care options as well as providing additional information, possibly about other sustainable consumer behaviours and initiatives. Some clothing designers and companies have developed their own low-impact care labels. Green Label's care label uses a combination of standardized and original care symbols along with conversationally written descriptions that encourage its customers to wash with eco-friendly detergents and line dry if possible or machine dry if they must (www.greenlabel.com).

Dombek-Keith (2009) proposed a 'green clothing care (GCC) label' (Figure 3.1.2) as a best-practice guide for reducing the environmental impact of clothing care. This label both complements and expands current care labels by selecting low-impact laundering options from existing care symbols, developing two new care symbols for when to wash and load size that address reducing laundering frequency, and extending the idea of care to repairing, reusing and recycling. The new care symbol designs are pictorial representations of their meanings, making them easy to understand. Just as the Energy Star label has brought consumer awareness to the efficiency of household appliances, designers could use the green clothing care labels on their clothing to increase consumer awareness and demand for clothing that can be cared for using sustainable methods. To demonstrate how the symbols could also be a promotional tool, Dombek-Keith designed a series of four GCC T-shirt logos that featured the new and current care symbols in combination with catchy slogans to explain their meanings. T-shirts and other clothing articles have often taken on the role of billboards, in this case educating and promoting consumer awareness of high- and low-impact clothing care behaviours while 'labelling' the consumer who wears the T-shirt as interested in sustainable clothing care.

Sometimes the best way to motivate people to change their consumer behaviours is to personalize education so they realize that their own actions have actual impacts and simple behaviour changes can make real differences. To accomplish this, Dombek-Keith developed the 'clothing care calculator' (Dombek-Keith and Loker, 2009), an interactive Excel-based tool that helps consumers evaluate the economic costs and environmental impacts (i.e. carbon emissions) of their typical laundering behaviours based on their laundry equipment, household energy costs and sources, and weekly load sizes and frequency. Along with the calculation, the program explains how these costs and impacts could be reduced through a variety of low-impact laundering choices and behaviours, such as switching to an Energy Star washing machine and line drying. This program and other similar calculators for laundry energy costs are available online.[1] Designers could encourage their customers to utilize such calculators or use them themselves to develop a variety of care scenarios highlighting different degrees of environmental impacts to include on garment promotional materials. Then consumers could compare these scenarios against their own laundering behaviours to gauge their probable environmental impacts.

Clothing designers and companies also have taken steps to educate themselves about how their products are impacting the environment. Many clothing companies including Patagonia, Timberland, Levi Strauss and Marks & Spencer have conducted life cycle assessments (LCA) of their products that evaluate the actual and potential environmental impacts at each life cycle stage, from design through disposal. Such assessments not only establish the designer or company's commitment to sustainability but, more importantly, provide them with valuable information to help determine new design and production choices that can reduce overall resource use. For many clothing companies, their LCA results have demonstrated the potential for huge energy savings through low-impact laundering behaviours, so they are actively promoting sustainable clothing care to consumers. Levi Strauss (2009) is asking its customers to 'wear responsibly' by laundering less frequently, washing in cold, line drying and donating used jeans for recycling. As part of Marks & Spencer's 'Plan A' initiative of 180 commitments toward sustainability by 2015, it is asking people to pledge to lower their washing temperature to 30°C in order to reduce energy consumption.[2]

The clothing design itself can provoke people to question their consumer behaviours as well as express their environmental and social concerns. Eco-friendly designer and researcher Kate Fletcher is an advocate for designing clothing with low-impact care requirements, a goal she believes will also require consumers to re-evaluate their own and society's definitions of cleanliness as a part of proper hygiene. The objective of her 'No Wash' design in the '5 Ways' project was months of wear without laundering, aided by reducing odours with underarm ventilation and allowing stains to be wiped off or stay in place as a symbol of the wearer's commitment to sustainability (Fletcher, 2008, p87). The design may have pushed the boundaries of common acceptance, but perhaps some people realized that their standards

110

for cleanliness were just as extreme. Designer and founder of 'The Uniform Project' Sheena Matheiken is challenging others to rethink the size and versatility of their wardrobes by wearing one dress for 365 days, modifying it each day with additional layers and accessories from their wardrobe and those of friends (Figure 3.1.3). Her non-profit project benefits the Akanksha Foundation to raise funds for a school in India (Davis, 2009), demonstrating another socially responsible commitment. By designing statement pieces that recognizably promote such sustainable initiatives, designers can, as Fletcher advocates, 'create images of *what might be*' in order to start conservations that lead to positive changes in consumer behaviour (personal communication, 16 November 2007).

Designing Sustainable Clothing That Enables: Low-impact Care, Extended Use and Meaningful Relationships

While clothing care can demand significant resource use on its own, designers must consider the potential combined environmental impacts over all stages in a garment's life in order to achieve sustainable clothing consumption. This goal is very tricky as design choices that could reduce resource use in one life stage may increase it in another. For example, many eco-friendly designers currently focus on choosing sustainable materials that are non-toxic, renewable, biodegradable and produced using less energy, like organic cotton. Yet consumers who want their cotton clothing to be clean, white, fresh, soft and smooth probably use significant resources in the care stage for hot water, bleach, fabric softeners, tumble drying and ironing. In addition, the unpredictable nature of consumer behaviour can result in the same two garments having very different environmental impacts over their lifetimes depending upon how their respective owners use, launder and dispose of them. For example, Franklin Associates (1993) compared two different laundering scenarios for a polyester blouse: when washed in cold and line dried, the blouse used 78 per cent less energy over its assumed lifetime than when washed in warm and machine dried. While consumers are ultimately responsible for the environmental impacts of clothing use, designers can enable consumers to choose low-impact laundering options, extend the garment's useable life and develop meaningful relationships with their clothing, all of which could reduce the resources used in overall clothing consumption.

111

Designers can choose materials with special fabric finishes that prevent stains and repel odours so that consumers could launder less frequently. Typically such finishes are coatings applied to fabric surfaces that reduce in effectiveness after numerous washings and can pose environmental concerns. Teflon® has been applied to clothing items, like some Dockers® pants, to provide stain resistance. Yet it is a toxic, possibly carcinogenic chemical that does not degrade in the environment (Savan, 2007), and the coating gums up recycling machinery (Ward, 2006). Silver ions reduced to the nanoscale are being used for anti-microbial finishes to kill microbes that cause odours and spoilage in a wide range of products, from socks to house paints (Erickson, 2009). Environmental groups are generally concerned about nanoparticles (defined as 100 nanometres (nm) or less) as they are more reactive than their larger versions and more permeable to bodily tissues (Miller and Senjen, 2008), and with silver in particular as it is a heavy metal groundwater pollutant that is toxic when ingested, obtained by environmentally damaging mining and could lead to resistant bacterial strands (Patagonia, 2010; Woodbury, 2008). Instead of silver, clothing brand Patagonia uses its own natural anti-microbial / anti-odour coated finish called Gladiodor®, derived from crab shells that are food industry by-products, on its Capilene® baselayer clothing (Patagonia, 2010). Recently, nanostructural finishes have been developed that are permanent and do not rely on toxic particles. For example, Nano-Tex Resists Spills™ is a stain-resistant finish that applies 'whiskers' to the fabric surface forming a barrier that prevents liquids and oils from soaking into the fibres (Resists Spills, 2010). AEGIS EcoFresh® is an anti-microbial finish with a sword-like surface that

LEFT | Figure 3.1.3 The Uniform Project series. www.uniformproject.com

pierces the cell membranes of microbes to destroy them.[3] A different approach to odour fighting could be achieved by actually promoting the growth of certain microbes that could both clean clothes and emit pleasantly sweet smells. Clothes with these microbes would need to be fed rather than washed to keep them clean (Samuel, 2001). Perhaps such anti-stain and anti-odour finishes could make Fletcher's 'No Wash' design idea acceptable to consumers, at least for particular applications like outerwear and accessories. At the minimum, such anti-stain and anti-odour finishes help prevent damage caused by stains or mildew that could shorten clothing lifespan, but reduction in laundering frequency is still dependent on the individual owner's care practices.

Special finishes can also make garments requiring special energy-intensive and potentially toxic care, such as dry cleaning, easier to care for using more energy-efficient and non-toxic methods. Marks & Spencer has developed exclusive technologies to create machine-washable wool blend men's suits and updated versions that also can be machine dried and do not need ironing (Rohwedder, 2006). In addition, Konaka in Japan is offering the 'Shower Clean Suit' for men and women that could change clothing care lifestyles toward washing some clothing in the shower (The suit, 2008). These suits are made from specially constructed wool fabric that can be rinsed clean with just two to three minutes of warm water, possibly during the wearer's evening shower. Then they can drip dry overnight, ready to wear in the morning thanks to an anti-wrinkle finish made from a natural amino acid found in fingernails and hair. Compared to dry cleaning, home laundering offers cost savings, convenience and environmental benefits that are clear and desirable. However, wash-and-wear formal attire has historically been viewed as cheap and poor quality, so it may take time for men to adopt such new solutions (Rohwedder, 2006). Designers could help further consumer acceptance of wash-and-wear formal clothing by adapting these innovations into stylish, quality designs.

Despite the heavy impacts of the use stage, extending a garment's life through better quality and increased usability can result in fewer resources used in the overall clothing consumption system if new clothing production is reduced (Collins and Aumônier, 2002). The Strategies towards the Sustainable Household (SusHouse) Project (Bras-Klapwijk and Knot, 2001) analysed the potential benefits for reducing the environmental impact of clothing through four 'design orienting scenarios' (DOSs) that compared the benefits of outsourcing clothing cleaning and repair along with reducing clothing consumption through innovative wardrobe choices, including:

- owning fewer, high quality clothes;
- sharing a pool of clothing within a neighborhood;
- leasing clothes from a service;
- purchasing new or second-hand clothing from the internet.

Environmental impact assessments for 2050 showed that all four DOSs reduced overall environmental impacts due to decreased clothing consumption since garments had longer, more useful lives as a result of improved quality and maintenance along with limited or shared wardrobes. Garment designs that are made with quality materials and durable constructions can be low-impact cleaned and easily maintained, and transcend the fast-fashion trends enabling consumers to extend and enhance the wearability of their wardrobes. Designers can further facilitate consumers owning fewer clothes by offering rental or sharing options and organizing take-back programmes to give unwanted clothing a second usable life with a new owner and wearer.

Designing updatable clothing can help prevent garment lives from being cut short by outdated styles, ill fit, and wear and tear, as well as allow wearers to customize their looks, potentially resulting in greater satisfaction and stronger personal connections with their garments. Dombek-Keith's 'Suit Yourself' designs (2009) (Figure 3.1.4) alter in style with fashion trends or in fit as the wearer's body changes. They are well crafted from high quality, eco-friendly fabrics to provide durability and support sustainable production practices. Each design is composed of base pieces (i.e. jacket body and waistband yoke) and a collection of changeable components, including jacket collars and cuffs and skirt and pants bottoms. The changeable components attach and detach using a system of separating zippers and snaps that allow separate laundering of components where staining is more likely, style updates or replacement when worn out. Fitting features such as sliding buckles and lacings accommodate shifting waist sizes over time. In addition, personal engagement through changing and customizing the designs can foster a deeper sense of connection that could encourage consumers to care for their garments better and for longer. For a sustainable design student project, UK student Zoe Fletcher designed a similar modular clothing concept: a line of knit sleeve combinations and trims, which attach to a base garment or can be used as accessories, like a bag.[4] Such versatility of looks and uses can help reduce clothing consumption by making fewer basic garments that can be added to for renewed style, allowing for longer garment lives and greater consumer self-expression.

BELOW | Figure 3.1.4 'Suit Yourself' modular/updated suit designs. Dombek-Keith (2009)

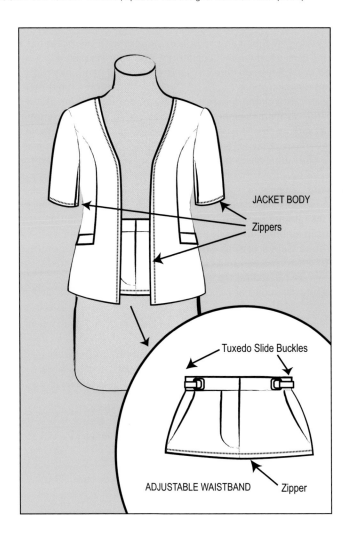

JACKET BODY

Zippers

Tuxedo Slide Buckles

ADJUSTABLE WAISTBAND   Zipper

Designing, or redesigning, garments to have multiple lives can also extend the useful lifetimes of garments. Chen and Lewis (2006) propose a model of sustainable design called 'five lives of a piece of cloth'. It starts with an uncut piece of cloth that can be used as a wrap, then transformed into its first cut-and-sewn design, a second re-cut and re-sewn design, another cut up and re-sewn item that is often an accessory or object other than garment, and the final stage where the fabric is ground up into compost for natural fibres or reuse or disposal for synthetic fibres. Fletcher (2005) created a similar design approach in her 'Nine Lives' design from the '5 Ways' project. Use of the garment through these multiple lives depends on appropriate care for freshening and stain removal and an appreciation of the attractiveness of patina from long wear that enhances the aesthetic of fabric through knowledge of its long life. Similarly, the entire life path of a garment can be re-imagined in ways that enhance its meaningfulness. Dombek-Keith's 'The Dowry Dress' (2009) is a wedding dress that represents the bride and shares in the life of the marriage through three features: recycling of special garments from the bride's past to add personal meaning, creating reusable components that can be worn again for future special occasions and refashioning the skirt of the dress into keepsakes that mark milestones of the marriage.

CONCLUSION

All of these design approaches can provide designers with the potential for making sustainable clothing care a fashion in its own right. Far from the past when sustainable clothing consisted of sack-like, un-dyed organic cotton and hemp basic apparel, we are proposing clothing designs where the aesthetic is strong and the function is sustainable. Designers can incorporate clothing care considerations at the design stage that reduce resources used in the clothing consumption system by enabling and promoting low-impact care behaviours as well as creating designs that are high quality, adaptable and foster personal meaning in ways that extend garment life. Clothing comes in many types for a variety of use, so one single sustainability solution is not feasible. Therefore designers should offer a set of options for consumers to choose from so that they can adopt a set of new clothing care behaviours, such as reduced laundering frequency and low-impact cleaning methods, smaller high quality wardrobes that are more versatile, and longer, more useful and meaningful garment lives with potentially multiple reincarnations. In the end, the garment can embody the environmentally responsible methods the designer used to create it and the consumer uses to care for it, becoming a symbol of the consumer's personal commitment to sustainability. Applying this expanded understanding of sustainable clothing care can reduce the impact of the overall clothing consumption system one designer and one consumer at a time.

NOTES

1. Green clothing care calculator available at www.laundrylist.org/why/calculator. Other online laundry calculators include www.yourhome.gov.au/tools/water-savings. html and www.energystar.gov/ia/business/bulk_purchasing/bpsavings_calc/ CalculatorConsumerClothesWasher.xls

2. Take Marks & Spencer's pledge to Keep clothes clean at 30° at: http://plana.marksandspencer.com/you-can-do/climate-change/15/

3. www.microbeshield.com

4. View Zoe Fletcher and other students' sustainable clothing designs at http:// fashioninganethicalindustry.org/studentwork

REFERENCES

Allwood, J. M., Laursen, S. E., Malvido de Rodriguez, C. and Bocken, N. M. P. (2006) *Well Dressed? The Present and Future Sustainability of Clothing and Textiles in the United Kingdom*. University of Cambridge Institute for Manufacturing, Cambridge, UK

Ashley, S. (1993) 'Out of the frying pan, into the dryer', *Mechanical Engineering-CIME*, vol 115, no 3, p120

Ashley, S. (1998). 'Energy-efficient appliances', *Mechanical Engineering*, vol 120, no 3, pp94–97

Bras-Klapwijk, R. M. and Knot, J. M. C. (2001) 'Strategic environmental assessment for sustainable households in 2050: Illustrated for clothing', *Sustainable Development*, vol 9, no 2, pp109–118

Cameron, B. A. (2007) 'Laundering in cold water: Detergent considerations for consumers', *Family and Consumer Sciences Research Journal*, vol 36, no 2, pp151–162

Chen, C. and Lewis, V. D. (2006) 'The life of a piece of cloth: Developing garments into a sustainable service system', *International Journal of Environmental, Cultural, Economic and Social Sustainability*, vol 2, no 1, pp197–207

Collins, M. and Aumônier, S. (2002) *Streamlined Life Cycle Assessment of Two Marks & Spencer plc Apparel Products*, Ref 7815, Environmental Resources Management, Oxford

Davis, S. (2009). 'Style with substance: Q&A with the founder of The Uniform Project', *The Daily Green*. www.thedailygreen.com/environmental-news/latest/uniform-project-interview-081410 (accessed 1 February 2010)

Dombek-Keith, K. (2009) *Re-fashioning the Future: Eco-friendly Apparel Design*. VDM, Saarbrücken

Dombek-Keith, K. and Loker, S. (2009) 'Clothing care calculator: An interactive tool to evaluate environmental impact', in L. Parker and M. Dickson (eds), *Sustainable Fashion: A Handbook for Educators*, Woodhead, Cambridge, UK, pp38–39

Dos Santos, A. (2007). 'Green "dry" cleaning', *Real Green*. www.greenamericatoday.org/pubs/realgreen/articles/drycleaning.cfm (accessed 1 February 2010)

Dunn, C. (2006) 'Sanyo's Aqua: Wash clothes without water', *Treehugger*. www.treehugger.com/files/2006/02/sanyos_aqua_was.php (accessed 29 March 2010)

Erickson, B. (2009). 'Nanosilver pesticides: EPA addresses data gaps, prepares to register more products', *Chemical & Engineering News*, vol 87, no 48, pp25–26

Federal Trade Commission (1995) 'FTC harmonizes care label regulations with Canada, Mexico'. www.ftc.gov/opa/1995/11/clab.shtm (accessed 1 February 2010)

Fletcher, K. (2005) 'The ecology of clothing', *Fourth Door Review*, vol 7, pp67–71

Fletcher, K. (2008) *Sustainable Fashion and Textiles: Design journeys*. Earthscan, London

Flex Your Power (2010a) 'Clothes dryers'. www.fypower.org/res/tools/products_results.html?id=100144 (accessed 29 January 2010)

Flex Your Power (2010b) 'Clothes washers'. www.fypower.org/res/tools/products_results.html?id=100122 (accessed 29 January 2010)

Franklin Associates (1993) *Resource and Environmental Profile Analysis of Manufactured Apparel Product: Woman's Knit Polyester Blouse*. American Fiber Manufacturers Association, Washington, DC

Frydendall, E. (2010) 'How laundry detergent works', TLC. http://home.howstuffworks.com/laundry-detergent.htm (accessed 29 January 2010)

Gerling, J. F. (2003) 'Microwave clothes drying: On the verge of commercial reality'. www.2450mhz.com/PDF/TechRef/Mw%20Clothes%20Drying.pdf (accessed 29 January 2010)

Jackson, T. and Michaelis, L. (2003) *Policies for Sustainable Consumption*. Sustainable Development Commission, London

Levi Strauss & Co. (2009) *Wear Responsibly*. www.levistrauss.com/Citizenship/Environment/WearResponsibly.aspx (accessed 7 March 2010)

Levinson, M. I. (1999) 'Rinse-added fabric softener technology at the close of the twentieth century', *Journal of Surfactants and Detergents*, vol 2, no 2, pp223–235

Lilley, D., Lofthouse, V. and Bhamra, T. (2005) 'Investigating product driven sustainable use.' Paper presented at

Sustainable Innovation 05: Global 'state of the art' in sustainable product/service development and design, 10th International Conference, Farnham Castle International Briefing and Conference Centre, UK

Mackenzie, D. (1997) *Green Design: Design for the Environment*. Laurence King, London

McCoy, M. (2003) 'Soaps and detergents: Whether for washing clothes or cleaning dishes, new products are increasingly the result of collaborative chemical development', *Chemical and Engineering News*, vol 81, no 3, pp15–22

Mediacenter, The (2008) *Business Overview on Dry Cleaners*. Weehawken, NJ

Miller, G. and Senjen, R. (2008) *Out of the Laboratory and on to Our Plates: Nanotechnology in Food and Agriculture*. Report for Friends of the Earth Australia, Europe and USA. www.foeeurope.org/activities/nanotechnology/index. htm (accessed 2 February 2010)

Morris, M. A., Prato, H. H. and White, N. L. (1984) 'Line-dried vs. machine-dried fabrics: Comparison of appearance, hand, and consumer acceptance', *Home Economics Research Journal*, vol 13, no 1, pp27–35

Palandre, L. and Clodic, D. (2003) 'Comparison of heat pump dryer and mechanical steam compression dryer'. Paper presented at International Congress of Refrigeration 2003, Washington, DC

Patagonia (2010). 'Technology: Gladiodor® natural odor control'. www.patagonia.com/web/us/patagonia.go?slc=en_US&sct=US&assetid=10161 (accessed 23 January 2010)

Petkewich, R. (2005) 'Cold-water laundry detergent is a hot idea', *Environmental Science and Technology A-Page*, vol 39, p478A

Project Laundry List (2007) 'Top reasons to hang out your clothes'. www.laundrylist.org/index2.htm (accessed 28 March 2008)

Resists Spills (2010) Frequently Asked Questions, http://www.nano-tex.com/applications/apparel_P1_FAQ.html (accessed 28 March 2008)

Rohwedder, C. (2006) 'Machine-washable suits make traditionalists recoil', *The Wall Street Journal*. www.post-gazette.com/pg/06110/683735-314.stm (accessed 15 February 2010)

Samuel, E. (2001) 'No more laundry', *New Scientist*, vol 171, no 2298, p25. Gale Group database (accessed 19 January 2010)

Savan, L. (2007) 'Teflon is forever', *Mother Jones*. http://motherjones.com/environment/2007/05/teflon-forever (accessed 29 March 2010)

Shove, E. (2003) 'Converging conventions of comfort, cleanliness and convenience', *Journal of Consumer Policy*, vol 26, no 4, pp395–418.

Shove, E. (2004) 'Sustainability, system innovation and the laundry', in B. Elzen, F. W. Geels and K. Green (eds), *System Innovation and the Transition to Sustainability: Theory, Evidence and Policy*. Edward Elgar, Northampton, MA, pp76–94

Steinemann, A. C. (2009) 'Fragranced consumer products and undisclosed ingredients', *Environmental Impact Assessment Review*, vol 29, pp32–38

The suit (2008). 'The suit you can wash in the shower'. http://web-japan.org/trends/07_lifestyle/lif080707.html (accessed 29 March 2010)

UN (United Nations Division for Sustainable Development) (2004) 'Changing consumption patterns' (Agenda 21: Chapter 4). www.un.org/esa/sustdev/documents/agenda21/english/agenda21chapter4.htm (accessed 15 December 2007)

US DOE (US Department of Energy) (2005) 'Table HC6.10 home appliances usage indicators by number of household members'. Energy Information Administration, Office of Energy Markets and End Use, Forms EIA-457 A, B, C, of the 2005 Residential Energy Consumption Survey. www.eia.doe.gov/emeu/recs/recs2005/hc2005_tables/detailed_tables2005.html (accessed 29 January 2010)

US EPA (US Environmental Protection Agency) (2010) 'Considerations for partnership in Design for the Environment (DfE) formulator initiative'. www.epa.gov/opptintr/dfe/pubs/formulat/consider/index.htm (accessed 3 February 2010)

Ward, G. (2006) 'Fashion, affluence, and recycling'. Presented at Green Chemistry Conference 2006: Green Solutions and Sustainability in Textiles and Fashion, University of Leeds, UK

Woodbury, M. (2008) 'Will bacteria develop resistance?' *Los Angles Times*. http://articles.latimes.com/2008/aug/04/health/he-nanoside4 (accessed 23 January 2010)

Workman, J. E. and Choi, Y. (1999) 'Consumers' understanding of care label symbols', *Journal of Family & Consumer Sciences*, vol 91, no 4, pp63–70

3.2 CASE STUDY
SLOWING FASHION

Dr Gene Sherman, Executive Director of the Sherman Contemporary Art Foundation in Sydney, Australia, has been a collector of Japanese fashion for over two decades. A carefully considered acquisition and 'retirement' process means that at any given time she has approximately 25 outfits in her wardrobe. For Gene the mantra 'buy less, buy better' is very much in evidence; the pieces are high-quality designer ready-to-wear from Yohji Yamamoto, Issey Miyake and Comme des Garçons by Rei Kawakubo. The work of these Japanese designers can be described as timeless, original in design and often unusual, which allows the pieces to easily transcend a six-month fashion season.

Where for many of us fashion shopping can be a casual and relatively regular activity, for Gene it is a carefully planned exercise. Gene dedicates half a day to shopping during her two annual trips to Tokyo and when she acquires a new piece, she 'retires' an existing piece from within her wardrobe. The retired piece is photographed and details are recorded such as date and place of purchase and where the garment was worn. The garments are then packed in conservation boxes with the archival information. Gene Sherman has donated a collection of the retired garments to the Powerhouse Museum in Sydney, which were subsequently exhibited and the garment stories are recounted through the museum's online catalogue. Gene's documentation of her retired pieces demonstrates that clothes can be a source of memories, linking us to a place and a time. These connections can result in a diminished need for buying excessive clothes; in this case the satisfaction normally gained through the act of buying a new garment can be matched through a reconnection with an existing garment. For Gene, this slower pattern of consumption enables her to create a deeper level of engagement with the clothes themselves and to build over time a personal ongoing narrative and set of associative values.

Clothes fulfill different kinds of material and non-material needs and to Gene her clothes are a source of pleasure. Fashion to Gene is a reflection of her aesthetic, intellectual and emotional needs and she seeks these responses whether she is buying artworks or fashion as wearable art. Most of us own garments that we have cherished and treasured for years but perhaps we should begin to question the responses a garment may elicit from us. What are our emotional attachments to particular pieces of clothing, and why do clothes connect with us emotionally? The responses that we gain can in turn inform further fashion acquisitions of clothes. But an engaging garment does not have to be an expensive designer piece. While Gene's approach to the editing and documenting of her wardrobe admittedly shows more discipline than an average fashion consumer, a smaller carefully considered wardrobe can provide the wearer with the opportunity to engage with clothing on a greater personal level.

The online collection is available at:
www.powerhousemuseum.com/exhibitions/contemporary_japanese_fashion.php

LEFT | Figure 3.2.1 Portrait of Dr Gene Sherman. Photographer James Mills for Harpers Bazaar

ABOVE | Figure 3.2.2 As worn by Gene Sherman, Issey Miyake 'PLEATS PLEASE', pleated polyester jacket, Autumn/ Winter 1999. Collection: Powerhouse Museum, Sydney. Photo: Sotha Bourn. Donated through the Australian Government's Cultural Gifts Program by Dr Gene Sherman, 2009

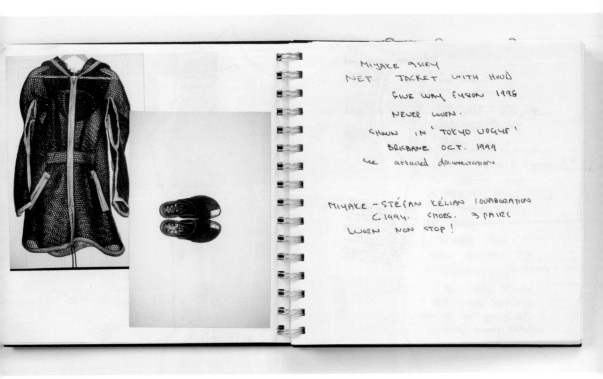

ABOVE | Figure 3.2.3 Gene Sherman's collection notebook, Sydney, Australia 1999. Collection: Powerhouse Museum. Photo: Chris Brothers. The notebook includes Gene's comments and photographs about the items in her fashion collection.

122

3.3 CASE STUDY
PERSONALIZING FASHION

Can the consumer inspire a counter-culture movement through creative acts of intervention and does this signal the power of the individual in shaping our future fashion industry? In 2005 Alex Martin, a performance artist based in Seattle, US, set out to wear the same dress for a year in a project she documented online as The Little Brown Dress. She made two identical dresses from brown cotton twill although she eventually wore only one. She washed the dress regularly, layering it with different garments each day, as much for appearance as for climate. Martin's aim was to confront the drive for over-consumption in fashion, particularly targeted towards women. She replaced buttons as they broke, and only changed garments for sleeping and swimming.

Martin has suggested the project might be called the 'intentional wardrobe' or a 'fashion de-tox diet'. Much can be learned from the project that fashion designers can incorporate in the design and production of clothing. Versatility was key with the brown dress, and it provides a key consideration for fashion designers who want to ensure a long use-life for a garment. Designers can further slow the typical fashion cycle by introducing trans-seasonal collections.

From the consumer's point of view, Martin's project questions the types of relationships we have with our clothes, presenting an alternative mode of fashion consumption. Through the year Martin documented each day on her blog with photographs. Reflecting on the year, Martin asked: 'Did I look crazy? Most people didn't even notice that I was always wearing the same dress day after day.'

In the Fashioning Now exhibition, 365 photographs showed the diversity of looks Martin was able to create in the course of a year with one dress. These showed that change in appearance – change in fashion – does not need an entire change of wardrobe. Change is part of fashion, but it is often from the false pace of fast change perpetuated by the industry that fashion's problems with sustainability arise. Though Martin describes the project as 'a one-woman show against fashion', it should serve the industry well in highlighting the value of producing durable, versatile pieces.

More recently, from May 2009 to May 2010, New York-based Sheena Matheiken echoed Martin's endeavour for The Uniform Project, raising over US$100,000 for the Akanksha Foundation, a charity that helps provide schooling for children in India. For a year Matheiken wore a black dress she designed with Eliza Starbuck. Other garments and accessories came from Etsy and eBay, while many designers donated these for Matheiken to wear. In yet another project that questioned consumerist fashion, from 1991 to 2002 Andrea Zittel, an American artist, would make a smock dress and then wear it every day for six months, repeating the process for each 'season'. Projects like these, as well as Dr Gene Sherman's approach to an edited wardrobe discussed elsewhere in the book, serve to remind us that being fashionable is not reliant on vast quantities of clothes that tend to clutter our wardrobes.

LEFT | Figure 3.3.1 The Brown dress project series (2005)

OVER | Figure 3.3.2 For 365 days Alex Martin wore the same brown dress

TIMO RISSANEN

3.4 DESIGNING ENDURANCE

This paper examines the relationship between fabric that is wasted through conventional fashion design and manufacture, and the physical and visual ability of a garment to endure. Research shows that designing garments without creating fabric waste in the process is feasible (Rissanen, 2008), and that design strategies exist to incorporate fabric that is normally wasted 'back' into a garment. The need for garment repair can be delayed, and repair as well as later alteration of the garment can be made easier. Strategies that venture beyond merely designing classics also exist that can make the garment more visually durable. Considered fashion design can positively impact on extending the useful life of a fashion garment. In a broader sense, this paper supports a shift from sustainable design to design for sustainability, calling for design that fosters more sustainable lifestyles instead of 'sustainable' products.

The current disposability of clothing is problematic. While clothes are seemingly durable goods, they are often marketed as fast changing fashions and made of increasingly inexpensive materials with quality of construction often neglected. Clothing today is more economically accessible than ever before, at least in developed nations (Allwood et al, 2006, pp11–12). Compromises in the quality of fabric and construction, and the outsourcing of manufacture to nations of cheap labour enable us to buy clothing in quantities not seen before. Despite the relatively low economic costs of fabric and clothing, these should be treated as precious and valuable due to the investments embodied by the fabrics and garments, investments made during fibre generation, design and manufacture. These investments may include water, material resources, energy and time. Fabric is a product with an ecological footprint attached but arguably not always treated as such by the fashion industry, and nevertheless relatively small increases in material inputs (fabric) at the fashion design stage can result in more physically and visually durable garments. Hypothetically this could result in a lesser need for new clothes.

Through a review of texts on garment construction, repair and maintenance, and case studies of men's shirts from the 18th century to today, it becomes clear that various transformative acts by the consumer pertaining to garment maintenance can become a fashion design consideration, alongside aesthetics, economics and ergonomics. The fashion designer cannot entirely control what happens to a garment in the consumer's hands. With the pattern maker the designer does, however, determine the degree to which the garment may be physically transformed after its purchase without adding any new fabric to it. Types of transformation within the garment become possible based on decisions and actions by the fashion designer and the pattern maker. These decisions can help delay the need for transformation by making the garment more physically and visually durable. Transformative practices covered in this paper are repair and alteration, which can prolong the useful life of a garment. Laundering by the consumer contributes significantly to the ecological footprint of a garment, and it may seem that the fashion designer can in no way control this. Based on the review, this paper proposes that some links may exist between the structural design of a garment and a reduced need for laundering, a view reinforced by Dombek-Keith and Loker elsewhere in this book. Following the case studies, the paper discusses a contemporary fashion garment, the design and make of which have been informed by the case studies. The paper will conclude with a call for a shift from sustainable design to design for sustainability in fashion design: design that fosters longer, less impactful lives for the garments we own.

TRANSFORMATIVE PRACTICES, FABRIC WASTE AND FASHION DESIGN

In the course of a few generations our relationship with clothes has changed somewhat, alongside the quantities of clothes we own. For most of our grandparents, mending and altering clothes was part and parcel of owning clothing. A lack of skills may be a contributing factor, but for our generation it is simply easier and relatively cheap to replace a damaged garment (Allwood et al, 2006, p39). Furthermore, why repair something that will be out-of-date soon anyway?

Before a garment is cut, a marker is made for it; a marker contains all the garment pieces. To improve durability through an increase in fabric within a garment, it may be unnecessary to use more fabric, and thus increase garment cost, than what the marker already contains. Fifteen per cent of the fabric in the marker of an average garment is wasted (Cooklin, 1997, p9); this waste is disposed of after cutting. This may not sound substantial until one thinks of the entire global fashion industry – 15 per cent of all fabric used by it is wasted. As some of the case studies show, much of this 'waste' could be incorporated into garments, for example, to reinforce parts of a garment that are prone to stress. To avoid using more fabric than usual to improve durability, the fashion designer needs to be aware of the garment's patterns and its possible marker compositions throughout the entire design process. To avoid increasing the total amount of fabric used, the reinforcement pieces need to come from 'gaps' between the main garment pieces in the marker.

Where can the wasted fabric be used in a garment? Drawing from texts on garment care as well as construction, this potential is now investigated. *The Wear and Care of Clothing* by Guilfoyle Williams (1945) is a guide for the consumer on choosing clothes that fit and how to care for them for maximum use-life. The book has a preventative focus. Repair is barely discussed; instead the author focuses on how to delay the need for repair. Although aimed at the consumer and despite its age, the book remains a valuable resource for fashion design and pattern-making. Interesting is the concern for quality of construction in inexpensive garments, and an entire chapter is devoted to identifying durable construction. The relationship between the amount of fabric in the 'turnings' (hems and seam allowances) and durability is discussed in some detail (Williams, 1945, p55).

Drawing from this, some of the 'wasted' fabric may be incorporated into larger allowances, which enable alteration, whether to accommodate a change in the wearer or fashion. In mass production, the machinist uses the seam allowance as a visual guide for sewing. Therefore seam allowances are almost always a uniform width, rarely exceeding 15mm. Usually the two pieces to be joined also have the same seam allowance; the machinist can visually align the pieces easily while sewing. On the contrary, clothes created through the more labour-intensive haute couture and bespoke tailoring practices often have wider seam allowances, and stitching lines are marked using thread or chalk (Shaeffer, 2001, pp38–39; Cabrera and Flaherty Meyers, 1983, pp58–61). In bespoke tailoring, the two seam allowances on the inside leg seam of a trouser may be two different widths, the wider back seam enabling future alteration to make the trouser thigh bigger (Cabrera and Flaherty Meyers, 1983, p16). The back crotch seam has a wide allowance at the waist, tapering to a narrower seam at crotch level, to allow for making the waist bigger if required (p229). Marking seam allowance widths with thread or chalk may not be cost-effective in mass production, but making stitching templates from card for the machinist seems practical. These are already used when usual marking methods are not possible. For example, a drill hole to indicate pocket placement is not appropriate on some fabrics, as the hole will fray during laundering. Instead, the machinist may be given a placement template made from card for correct pocket positioning. Similar templates could be developed for variable seam allowance widths.

Crown (1977) offers a manual for making new clothes out of old. While some featured projects involve taking apart a garment to make a new one, some of the projects alter a garment to fit the wearer or a new fashion. For example, to make a pair of pants larger, Crown (p42) suggests inserting gussets of fabric at the back waist, and taking this fabric from the hem or inside pocket. The wider the hem, the easier this alteration would be. If the pant had a seam allowance like that on tailored trousers described above, this insertion might not be necessary. In fact, many of the included projects would similarly be made easier with strategically planned wider seam and hem allowances. Perhaps 'future alterability' should become a consideration for fashion design. One obstacle for this may be the perception of seam allowances as waste. Some 5.5 per cent of the total fabric in a garment is in the seam and hem allowances, and according to Cooklin (1997, p10) the pattern maker is responsible for ensuring 'that all these allowances are the practical minimum possible'. This does not, however, account for the entire use-life of a garment; seam allowances in fact can be an asset, an investment in a garment's future. Allwood et al (2006, pp38–39) suggest that 'fast fashion' may be more difficult to repair for practical, economic and psychological reasons, perhaps tempering the view of seam and hem allowances as waste.

129

Two older manuals on domestic clothing construction discuss repair in detail. Both suggest that repair should be invisible. The chapter on mending in the older text reminds the reader several times that the end result should be inconspicuous (Woman's Institute Library of Dressmaking, 1923, pp151–157). Forty years later, another text notes: 'We no longer mend and darn the way our grandmothers did; the emphasis on what we mend has changed' (*McCall's Sewing in Colour*, 1964, p262); a chapter is dedicated to 'Repairing, remodelling, remaking'. The aim is the same: to make the mend invisible (p265). A visible repair somehow disturbs the look of a garment, devalues it. Could visible repair, and reparability, be an inherent aspect of a garment's appearance?

A look into dress history outside the Western tradition, to a Japanese island at the beginning of the 20th century, reveals an alternative: garments that constantly change in appearance yet remain essentially the same; garments that can remain in use indefinitely. The fishermen's coats from Awaji Island (Sadako Takeda and Roberts, 2001) are made of indigo-dyed cotton covered with white sashiko quilting stitches (Figure 3.4.1).

These coats gradually fade with washing and exposure. A spectacular aspect of these coats is their ability to 'absorb' repair without compromise to their aesthetic appeal. A hole is covered

130

ABOVE | Figure 3.4.1 Fisherman's coat from Awaji Island from the collection of the Hokudan Town Historical and Ethnographic Museum, Japan. Photography: Don Cole at Fowler Museum at UCLA

with a quilted patch of fabric; the patch is initially darker but fades over time. While the number of patches grows, the overall look of the coat is maintained: the patches become the coat. The coats may no longer be worn as the culture in which they originated has dramatically changed but they teach a powerful lesson. There is virtually no limit to how long such a garment could remain in use, living with its wearer(s). How could one apply this idea to contemporary fashion design? How could one design garments whose value is not diminished by alteration or repair? Perhaps this is not a problem solely for fashion design: the consumer would need to see beauty in the patina of the used, the repaired, for such practice to become widespread. Could the fashion designer contribute to creating opportunities for this beauty, in a culture distinguished by a desire for newness and novelty?

Fletcher (2008, p187) notes how clothes are presented to consumers as almost sacrosanct, resulting in passive fashion that discourages alteration and customization. Perhaps the key for fashion design is to embrace disturbance. For example, clothes by British label Junky Styling are made from second-hand clothes, and the two designers, Kerry Seager and Annika Sanders, also refashion clothes that customers bring in. Given that these clothes are already cut into, doing so again in the future is perhaps easier. New York-based fashion designer Tara St James (2010) discusses a fear of failure in a design context, noting that once the first cut – a disturbance – has occurred in the fabric, the fear dissipates: 'Once the fabric is cut it's no longer perfect, no longer pristine, no longer untouched, and can both mentally and literally be transformed.' This idea will be revisited in the context of creating potential for visual endurance.

HISTORICAL MEN'S SHIRTS: CUT, MAKE AND ENDURANCE

By treating shirts from a period spanning over two centuries under one heading, it is not the intention to belittle the work of dress historians who may be able to date a garment to a decade, even a year, based on its details. Rather, the aim is to find common features pertinent to the issues at hand. First this section examines how fabric has been used in shirts in the past to increase their physical durability, followed by a brief look at men's shirts today.

A man's linen shirt from England or America featured by Baumgarten et al (1999, pp105–108) dates from 1775–1790. The front and back body is one rectangle with the shoulders on the fold. The left side of the shirt has the original selvedge of the fabric as the seam finish. A t-shaped slash creates an opening for the neck. Two small fabric gussets are folded into triangles and inserted into the shoulder parts of this slash to achieve a sloping shoulder. The remaining gusset edges are gathered into the collar. Square gussets are also used under the arms to reduce strain. Each shoulder has a reinforcement piece sewn to the inside to strengthen the armhole where the sleeve joins the body. The shoulder reinforcements are not cut from the same fabric as the main shirt; they are from a coarser and possibly less expensive linen fabric. The entire shirt is made of rectangles and squares. It can be cut with very little waste, as the straight edges of the pieces can abut against each other for cutting. Because only one selvage remains, it is not possible to ascertain the width of the fabric and thus the amount of waste, but it is safe to say it would be below today's average. The authors note (p108) that the cutting layout of the pieces rarely changed; if a smaller or larger shirt was desired, a different width of fabric was used. Notably this shirt was altered, probably between 1810 and 1820: the collar and cuffs were replaced with similar but finer linen and the cuff width suggests this date. Furthermore, the original sewing thread was linen but the shirt was repaired with cotton thread, only developed in the early 19th century. Based on these dates, the shirt was in use for 20–45 years. Even the shorter estimate is arguably a longer time than what we would expect of a shirt now, and possibly demonstrates how precious textile products were then.

A reprint by Shep and Cariou (1999, pp24–25) of the English *Workwoman's Guide* from 1837 includes making instructions for a similar shirt. The drawing of the shirt is not complete, and the cutting diagram is difficult to relate back to the construction instructions. For example, the instructions refer to 19 separate pieces while the cutting diagram has 15 (the main body, a rectangle, is not included in the diagram). The remarkable feature of the cutting diagram is that all the fabric seems to have been used, and the sizes of some smaller gussets seem to have been determined by the amount of fabric left over by the larger pieces. This shirt is more complex than the 18th century shirt in that a greater number of gussets allow a further refinement of an otherwise rigid geometric garment shape. From the 1840s these 'square-cut' shirts were gradually replaced by 'tailored' shirts, which had curved armholes, sleeve heads and neck openings replacing the gussets. Shep and Cariou (pp5–6) attribute this shift partly to comfort and durability: the square-cut shirt accurately reflected neither the body nor the garment worn over it, subjecting it to more strain than a tailored shirt. The new tailored shirt necessitated the use of a pattern, while the square-cut shirt could be cut based on a cutting diagram and a set of measurements. The square-cut shirt did not disappear suddenly; its geometric elements can be seen in some shirts from the '"Keystone" Shirt System' of 1895, also among the reprints (pp120–123).

The tailored shirt of the 1840s in many ways resembles a simple man's shirt of the 21st century. In the most basic form, a contemporary shirt is made of two fronts, back, shoulder yoke (cut double), sleeves, sleeve plackets and cuffs (double), and a collar and a collar stand, both double. It might also have a breast pocket. The armholes, sleeve head and neck are curved and the side seams have varying degrees of waist shaping in them. When all the shirt pieces are placed on the marker, about 85 per cent of the total fabric becomes the shirt. What is to stop the fashion designer and the pattern maker from examining the marker, and investigating the possibility of incorporating some of the wasted fabric into the shirt? For example, part of the waste could become a reinforcement patch for the elbow area, while another could become a gusset at the side seam hem to reinforce a weak point. Such an investigation and redesign could take place with any type of garment. In a shirt, the possibilities for lengthening the garment's life are numerous. While the double yoke gives added support to the shoulder and neck area, the underarm could be further strengthened with another piece. It could be worthwhile to examine how much more fabric, if any, the shirt would use if a second collar, collar stand and a pair of cuffs were made. Crown (1977, p6) points out that these tend to be the first part of a shirt to show signs of wear, with the rest of the shirt usually relatively pristine; yet they are easily removed and replaced. Until the early 20th century shirts commonly had detachable collars, which allowed the separate, more frequent laundering of the collar, as well the possibility of varying the style (Roetzel, 1999, p50), providing variation for the wearer.

DESIGNING ENDURANCE

In a sociological account of the history of waste making in the US, Strasser (1999, pp38–52) provides an extensive section on clothes repair and second-hand clothes. Whether intentionally or not, Strasser delivers a richly detailed catalogue of garment care and transformation. Many examples are from 19th century 'housekeeping' manuals directed at women; significantly, they reveal where garments tend to wear out first. This information is crucial if fashion design is to address repair. Pedersen (2007) and Lockren (2010) have researched the repair and alteration practices within the wardrobes of individual women, while Palmer (2001) has investigated repair and alteration of Parisian haute couture among Canadian consumers in the 1950s. In brief, garments that have 'excess' fabric in them from the outset are easier to repair or alter – transform – later on. Fashion design and pattern-making can enable and facilitate these practices.

It may seem that zero-waste fashion design (fashion creation without fabric waste) has only emerged in recent years as part of the latest wave of 'sustainable fashion'. It is evident in Dorothy Burnham's (1973) research, however, that garments have been designed and made with little or no waste for centuries. Whilst Burnham's research focused on historical garments, even in a contemporary design context such an approach to design has been evident for several decades in the works of Claire McCardell, Zandra Rhodes and Yeohlee Teng, among others (Rissanen, 2008).

The Endurance shirt by the author exhibited in Fashioning Now in 2009 and 2010 (Figure 3.4.2) is cut without fabric waste when two shirts are cut at the same time; the marker in Figure 3.4.3 demonstrates this.

The shirt addresses two areas where fashion designers can have an impact. On one hand, the aim is for this garment to create a connection between designer-maker and wearer through incorporating craft techniques into a contemporary garment. Many of the designer's thought processes are explicit in the garment, aiming to connect with the wearer. For example, to create a blouson at the back, the shirt has an internal waist stay more common in women's haute couture. With the act of buttoning the stay the blouson becomes instantly explicit, communicating the stay's rationale to the wearer. Using the fabric's selvedges as external seam finishes aims to communicate the zero-waste nature of the shirt to the wearer. On

132

RIGHT | Figure 3.4.2 Endurance shirt by Timo Rissanen, (2009). Silversalt Photography

Endurance shirt

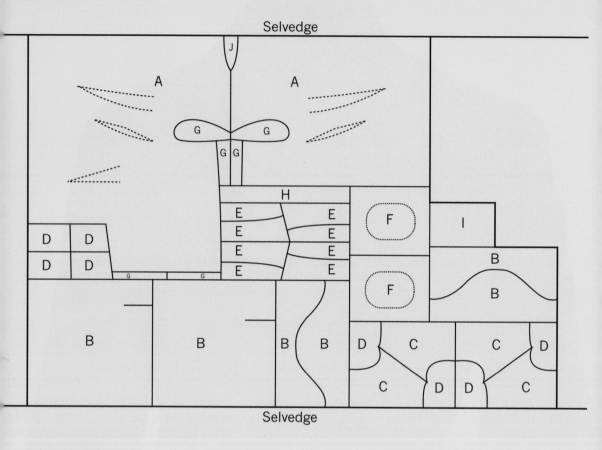

Selvedge

Selvedge

Fabric: 100% Linen
Fabric width: 135cm
Yield: 176cm

A: Body
B: Sleeve (including top sleeve lining)
C: Yoke
D: Cuff
E: Collar & stand
F: Elbow patch
G: Sleeve placket
H: Internal waist stay
I: Internal back pleat stay
J: CB Yoke appliqué

Dashed lines indicate darts and dart-tucks
Dotted lines indicate fold lines

the other hand, the shirt is designed and made to allow later alteration and repair. The hand quilting at back waist and in the elbow patches aims to suggest explicit mending. The objective of this is that later visible repair or alteration would not compromise the garments aesthetically. Furthermore, surplus fabric is designed into the garments to facilitate these activities. An internal patch at the back waist in the main fabric could be replaced with another fabric if the patch were required elsewhere in the garment. Furthermore, the elbow patches have 'excess' fabric folded underneath should it be required later.

THE FUTURE OF FASHION DESIGN?

Design for sustainability, as opposed to sustainable design, refers to design that fosters more sustainable behaviours in users. In the case of fashion garments, these behaviours include laundering and drying, repair and alteration, and delaying disposal, amongst others. Given the large impact often made by laundering and drying, these need further attention. Fashion designers should ask: can fashion design foster less intensive laundering and drying practices? The paper by Dombek-Keith and Loker in this book strongly suggests so. While much attention has been paid to fibre choice and reclaiming materials, the broader focus needs to shift to fashion design that supports more sustainable lifestyles among consumers.

An increase in the materials inputs, or incorporating more fabric into a garment at the design and manufacturing stage may result in a longer useful life for the garment: '...re-materialisation can be a dematerialisation strategy' (Tonkinwise, 2005, p27). Increased physical and visual durability can emerge from an increase in the amount of fabric used per garment, as well as an increased investment of time and energy in its construction. Hypothetically, greater durability could result in a lesser need for new clothes. With most garments currently produced through conventional mass-production methods, a considerable amount of fabric is wasted. This wasted fabric may, however, be incorporated back into the garment in various ways. The fashion designer and pattern maker may extend the use-life of a garment by using the wasted fabric to delay the need for repair but also enable or facilitate transformative practices in the future. Fabric waste and the resulting waste management costs can then be avoided. In some instances, the garment fabric may be used to give added body or structural support to a garment part, replacing a separate interlining fabric, often made of a different fibre. This could make the future disposal or recycling of the garment easier. Furthermore, transformative practices can invest a garment with added meaning for the wearer (Gregson and Crewe, 2003, pp163–167). For example, altering a garment to make it fit better may increase the time the consumer will retain it and care for it. Fashion design with pattern-making can enable or facilitate these transformative practices. Examining the marker can visualize for the fashion designer and pattern maker that not only is fabric wasted in an average garment, but that ingenuity and creativity offer potential for designing and making a physically stronger garment without increasing the total amount of fabric used per garment. Increasing a garment's visual durability may seem more of a design challenge. Fashion is often presented as fast changing yet most of us own garments we have had for a long time and still wear. Nevertheless, if alteration is required to update a garment or mending is needed to fix a damaged garment, how can the designer ensure that neither of these transformative practices will devalue a garment? How can we design garments that can age (patina), that can change and live with the wearer much like the coats from Awaji Island? In her paper in this book Fletcher discusses the Local Wisdom project; Edward's 'three stage jacket' is an example of such a garment in a contemporary context, having undergone three major transformations in its 40 years (and counting) of wear. It is proposed here that garments with an element of disturbance may be more easily visibly transformed. Outi Pyy of the DIY fashion blog Outsapop (www.outsapop.com) shreds striped Marimekko T-shirts for re-sale (Figure 3.4.4).

135

LEFT | Figure 3.4.3 Pattern layout for Endurance shirt, (2009)

Disturbing a design classic may make it easier for consumers to customize further. For 'undisturbed' clothes, there are 'cook books' by Otto von Busch: collections of methods of customization to empower consumers to make their clothes really theirs (Fletcher, 2008, pp195–197).

This paper has shown that historical dress can inform 21st century fashion design in the context of environmental sustainability. More ground could be covered. The 'Utility Scheme' that was in place in the UK during and after World War II (Sladen, 1995) placed various restrictions on clothing manufacture to make the most of scarce resources. The scheme's aesthetic restrictions may not be plausible in a contemporary design context, but the improvements in quality that the scheme achieved may hold a lesson for contemporary fashion design. Testament to this quality is a man's suit in the Victoria and Albert Museum in London, which the owner wore from 1945 to 1982 (37 years) (Hart, 1984, pp70–71). To achieve such a long use-life not only requires careful design and stringent garment manufacturing standards, but also a high quality fabric. The scheme shows that quality does not have to mean economic inaccessibility, something the contemporary fashion industry probably needs to investigate further.

Interestingly, some contemporary fashion designers, who show a new collection twice a year, do not equate showing seasonal collections with making obsolete the previous season's clothes. Issey Miyake puts it bluntly: 'To change every six months... is crazy. It's designer suicide' (Frankel, 2001, p48). A statement from the Maison Martin Margiela elaborates: 'We have always had garments that we continue to propose for many seasons in a row (in some instances twelve!). It remains more important for us that someone finds their way of dressing as opposed to a way of dressing as prescribed by anyone else or an over-riding trend' (Frankel, 2001, p35). Whether a link exists between originality in fashion design and a lesser degree of built-in obsolescence could be a topic for future research. To help designers extend the useful life of garments by design, Tonkinwise (2005, p27) sums up the design brief: 'Design timely things, things that can last longer by being able to change over time. Design things that are not finished, things that can keep on by keeping on being repaired and altered, things in motion.' Clothes that are never finished: a formidable yet an exciting task.

RIGHT | Figure 3.4.4 Shredded Marimekko T-shirt by Outi Pyy, (2010)

REFERENCES

Allwood, J. M., E. Laursen, S. Malvido de Rodríguez, C. and Pocken, N. M. P. (2006) *Well Dressed? The Present and Future Sustainability of Clothing and Textiles in the United Kingdom*. University of Cambridge, Cambridge, UK

Baumgarten, L., Watson, J. and Carr, F. (1999) *Costume Close-up. Clothing Construction and Pattern 1750–1790*. The Colonial Williamsburg Foundation in association with Quite Specific Media Group, Ltd., Williamsburg, VA and New York

Burnham, D. K. (1973) *Cut My Cote*. Royal Ontario Museum, Toronto

Cabrera, R. and Flaherty Meyers, P. (1983) *Classic Tailoring Techniques. A Construction Guide for Men's Wear*. Fairchild Publications, New York

Cooklin, G. (1997) *Garment Technology for Fashion Designers*. Blackwell Science, Oxford

Crown, F. (1977) *How to Recycle Old Clothes into New Fashions*. Prentice-Hall, Inc., Englewood Cliffs, NJ

Fletcher, K. (2008) *Sustainable Fashion and Textiles: Design Journeys*. Earthscan, London

Frankel, S. (2001) *Visionaries. Interviews with Fashion Designers*. V&A Publications, London

Gregson, N. and Crewe, L. (2003) *Second-hand Cultures*. Berg, Oxford and New York

Guilfoyle Williams, J. (1945) *The Wear and Care of Clothing. A Practical Guide to the Correct Choice and Care of Clothing*. The National Trade Press, Ltd, London

Hart, A. (1984) 'Men's dress', in N. Rothstein (ed.) *Four Hundred Years of Fashion*. Victoria & Albert Museum, London, pp49–74

Lockren, P. (2010) 'Can women's clothes from the 1900s inform the clothing of contemporary women?' 2010 IFFTI Conference: Fashion: Sustainability and Creativity, March 24–25, 2010. Fu Jen Catholic University

McCall's Sewing in Colour (1964) Paul Hamlyn, London

Palmer, A. (2001) *Couture and Commerce: The Transatlantic Fashion Trade in the 1950s*. University of Washington Press, Seattle, WA

Pedersen, K-A. (2007) 'Alteration in the clothes in one woman's wardrobe 1925–1940'. Dressing Rooms conference, 14–16 May 2007. Oslo University College, Oslo

Rissanen, T. (2008) 'Creating fashion without the creation of fabric waste', in J. Hethorn and C. Ulasewicz (eds), *Sustainable Fashion: Why Now? A Conversation about Issues, Practices, and Possibilities*. Fairchild Books, New York, pp184–206

Roetzel, B. (1999) *Gentleman. A Timeless Fashion*. Könemann, Cologne

Sadako Takeda, S. and Roberts, L. (2001) *Japanese Fishermen's Coats from Awaji Island*. UCLA Fowler Museum of Cultural History, Los Angeles

Shaeffer, C. B. (2001) *Couture Sewing Techniques*. The Taunton Press, Newtown

Shep, R. L. and Cariou, G. (1999) *Shirts and Men's Haberdashery: 1840s to 1920s*. R. L. Shep, Mendocino

Sladen, C. (1995) *The Conscription of Fashion: Utility Cloth, Clothing and Footwear 1941–1952*. Scolar Press, Aldershot

St James, T. (2010) 'Just cut', *The Square Project*. http://4equalsides.com/2010/04/28/just-cut/ (accessed 30 April 2010)

Strasser, S. (1999) *Waste and Want: A Social History of Trash*. Metropolitan Books, Henry Holt and Company, New York

Tonkinwise, C. (2005) 'Is design finished? Dematerialisation and changing things', in A-M. Willis (ed.), *Design Philosophy Papers. Collection Two*. Team D/E/S, Ravensbourne, pp20–30

Woman's Institute Library of Dressmaking (1923) Woman's Institute of Domestic Arts and Sciences, Scranton, PA

LAST
Chapter 4

LAST | INTRODUCTION

The closing chapter focuses on the end of life of fashion garments. When a consumer determines that a garment is no longer desirable for the purpose it was acquired for, it becomes textile waste. While the price of clothing has decreased, consumer spending on clothing has increased, resulting in significant increases in textile waste streams in the West (Allwood et al, 2006). As clothes become cheaper, it may be easier for a consumer to discard a garment and replace it with little consideration. A sustainable fashion industry of the future must identify ways of producing fashion that foster deeper engagements between wearer and garment, from point of acquisition through an appropriate, low-impact use phase to the eventual end of life of the garment. This will require new, closer relationships between the industry and fashion consumers. Alongside social innovation, technological advances will continue to bring about improvements in materials reclamation and recycling, leading to waste reduction.

Jana Hawley examines the complex global textile recycling industry. It is clear that reductions are needed in the levels of fashion consumption, but new strategies also need to be identified to divert clothing and textiles from landfill; as Hawley points out, textiles are nearly 100 per cent recyclable yet in the US much textile waste ends in landfill. Palmer (2001) argues that recycling must account as much for function as for material; arguably clothing is best recycled as clothes as the function of wearability is retained. Even when function may be lost, McDonough and Braungart (2002) argue that materials, including textile fibres, should be regarded as nutrients rather than waste at all times, shifting from a cradle-to-grave paradigm to cradle-to-cradle. While Hawley acknowledges that the complicated web of the textile recycling industry is not without its challenges, she identifies significant new business opportunities within the sector, showing the way forward in reducing textile waste.

Focusing on the role of user-makers in a post-growth economy of a sustainable future, Kate Fletcher draws from the Local Wisdom project, which centres on creative acts by ordinary consumers that somehow engage with sustainability in fashion. Fletcher argues that these acts, be they customization, sharing or not laundering a garment, infer an increase in the quality of the relationships people have with their clothes, leading to prolonged ownership, a reduction in waste, and perhaps most significantly, an improved quality of life. In a broader sense, design and use ought to be of the same whole, enriching each other creatively.

....

Allwood, J. M., E. Laursen, S. Malvido de Rodríguez, C. and Pocken, N. M. P. (2006) *Well Dressed? The Present and Future Sustainability of Clothing and Textiles in the United Kingdom*. University of Cambridge, Cambridge, UK

McDonough, W. and Braungart, M. (2002) *Cradle to Cradle: Remaking the Way We Make Things*. North Point Press, New York

Palmer, P. (2001) 'Recycling as universal resource policy', in C. N. Madu (ed.), *Handbook of Environmentally Conscious Manufacturing*. Kluwer Academic Publishers, Boston, Dordrecht and London, pp205–228

JANA M. HAWLEY

4.1 TEXTILE RECYCLING OPTIONS:
Exploring What Could Be

INTRODUCTION

The textile recycling industry is an age-old business that has been sorting, grading and conducting business in the same way for several generations. Clearly now is the time to consider new strategies that generate environmental and economic improvements that could catapult the industry to the next level and help rid us of the plethora of textile waste that is warehoused throughout the Western world. This chapter provides an overview of the industry and opens the door for critical discussions that need to be considered so that the industry can move forward to address the pending issues as consumer textile waste increases.

IMPACT OF FASHION ON RUBBISH

For the past 40 years, an increasing public awareness toward environmental issues has brought attention to the fashion industry – an industry that is fundamentally in opposition to the concept of environmentalism. As Brosdahl (2007) questions, clothing can be made sustainable, but fashion? In recent years, the fashion cycle has become shorter and shorter as fast fashion retailers such as Zara, H&M, Primark and Forever 21 sell clothing that is expected to be disposed of after being worn only a few times (Lee, 2007). As both industry and consumers continue to embrace the fast-fashion concept, the volume of goods to be disposed of or recycled has increased substantially. And most fast-fashion goods do not have the inherent quality to be considered as collectables for vintage or historic collections. Furthermore, as disposable income in countries such as China and India continues to increase, the supply chain of cheap textiles into the waste stream, including the critical problem of necessary disposal, will continue to escalate.

Over 12 million tonnes of textiles are sent to US landfill each year even though textile and clothing products are nearly 100 per cent recyclable (US EPA, 2008). The process of textile recycling is still not fully embraced and people do not completely understand the process. An additional purpose of this chapter is to explain about the textile recycling industry and to open the door for consideration for new value-added options and new consumer behaviour options.

OVER-CONSUMPTION CHALLENGES

A half century ago, in *The Waste Makers*, Packard (1963) warned about the shortcomings of post-war modern society. He argued that Americans had become a force-fed society with a vested interest in over-consumption, with no end in sight to the ever-increasing extravagant waste. Packard coined the phrases 'planned obsolescence' and the 'throw away society' and pointed to the significance of the social, economic and environmental implications of an unrestricted growth in consumption. While 50 years ago few paid attention to what seemed like an alarmist message, Packard's message now seems timely. Today, with considerations of overflowing recycling centres, and warehouses full of sorted and baled used clothing with no place to send it, over-consumption concerns are certainly looming. Furthermore, consumption patterns have spread dramatically to become a global issue. Massive populations in India and China are beginning to earn discretionary income – a right for every human on the planet, but also a concern that impacts on environmental issues.

Clothing is eventually cast off by the original consumer and enters the recycling stream. Depending on the consumer, clothing is thrown away, donated to charity or otherwise re-channelled for use or exchange value. Fashion marketers entice us to buy something new every season, sometimes with offerings that are truly new and exciting, but all too often the merchandise is simply a slight twist on last year's successful designs, offering the safe bet rather than taking a risk with shareholders' expectations. Consumers satisfy their whims, often overburdening their wardrobe space and probably their credit cards. The result is a clothing accumulation that stems from planned obsolescence, the very core of fashion and the epitome of what Packard was warning us about in the 1960s.

Elizabeth Wilson calls fashion 'dress in which the key feature is rapid and continual changes of style. Fashion... is change' (Wilson, 2003, p3). But regardless of how dynamic fashion is or how economically viable the fashion industry is, the global fashion industry is creating an overabundance of used clothing. The production of excess, inexpensive and often poor quality clothing depletes our resources (both natural and petroleum-based), contributes to the (ab)use of low-wage labour pools and creates a plethora of stuff to be siphoned through the waste management stream.

THE ENVIRONMENTAL CONCERN

According to the Environmental Protection Agency, the per capita daily disposal rate of solid waste in the US is approximately 1.95kg, up from 1.2kg in 1960. Although textiles seldom earn a category of their own in solid waste management data, the per capita consumption of fibre in the US is 38kg with over 18.14 kg per capita being discarded each year (US EPA, 2008). This discard rate is nearly three times greater than anywhere else in the world. When the environmental impact of the production of fashion is considered, particularly for items that are seldom or sometimes NEVER worn, the carbon footprint begins to become unsettling.

Almost 80 per cent of the textiles that go to landfill are petroleum-based synthetic fibres, while 20 per cent come from natural fibre sources (personal communication with a large

RIGHT | Figure 4.1.1 Baled used textiles (sweater)

ABOVE | Figure 4.1.2 Women cutting buttons off used sweaters to make blankets

US textile recycle company owner). While the natural fibres will theoretically decompose over time, landfill processes bury the textiles leaving little opportunity for even the natural fibres to decompose. Furthermore, textiles leave comparatively little toxic leachate or biogas during the decomposition process, yet they take space in a landfill and because textiles are nearly 100 per cent recyclable, there needs to be more attention to ways in which waste management and textile recycling companies can help divert textiles and clothing from the landfill (SMART, 2004).

THE TEXTILE RECYCLING PROCESS

Today's consumers are great at throwing stuff away, and replacing it with something new, maybe even replacing it with two things. But sometimes consumers want to see if they can reap a reward for their personal things by having a garage sale or taking gently used things to a consignment store. What are the other methods for getting rid of your stuff? While some municipalities have clothing and textiles in their list of recyclables, most do not. The relationship between rubbish and value is complicated, ambiguous and complex. Goods 'transfer and shift between and across cultural categories which are themselves fuzzy and striated: at one end it's rubbish, at the other end they have high [commercial and aesthetic] value' (Gregson and Crewe, 2003, p115). Many ways to part with things at the end of the personal use-life exist, but for most, the best solution is to take your used things to your favourite charity. Most charities will agree, though, that the supply far exceeds the demand. What the charity shops cannot sell in their local retail stores will then be sold by weight to the for-profit companies, commonly referred to a 'rag dealers'. Rag dealers sort for a wide variety of markets.

In work that is published elsewhere (Hawley, 2006, 2008, 2009) I fully explain how the textile recycling process is a system that requires a myriad of players, including:

- consumers who sell their cast-offs at garage sales or flea markets;
- charity shops such as Goodwill Industries and Salvation Army;
- for-profit recycling companies;
- manufacturers who develop value-added products from post-consumer waste;
- policy makers;
- developing countries that import bales of used clothing;
- 'pickers' who desire vintage collectables from the heaps of used goods (e.g. frayed Grateful Dead T-shirts or perfectly ripped pairs of faded Levis).

Textile recycling can be classified as either pre-consumer or post-consumer waste; a textile recycling company removes this waste from the waste stream and recycles it back into the market (both industrial and end-consumer). Pre-consumer waste consists of by-product materials from the textile, fibre and cotton industries that are re-manufactured for the automotive, aeronautic, home building, furniture, mattress, coarse yarn, home furnishings, paper, apparel and other industries. Post-consumer waste has been purchased and worn (or not – many things that go to the Goodwill with retail price tags still on them) by the consumer.

THE WORLD OF SECOND-HAND CLOTHING

Of all the textiles that are recovered, clothing makes up almost three-quarters of the content (SMART, 2004). What enters the used clothing market is quite different from what ends up on the rag dealer's conveyor belts to be sorted for other value-added purposes. Some consumers have a difficult time parting with what they view as treasured possessions – things never to be thrown away. Other items have a very short life (purchase mistakes, unwanted gifts or poor quality goods with expected short lifespans). The boundaries between what is treasured and what should be discarded are fuzzy. And the boundaries between what should be sold in charity retail shops and what should be sold to rag dealers are even fuzzier. Sourcing, sorting and disposal strategies vary by country, consumer, charity organization and rag dealer. Interactions that cause variability depend on policies and laws, connectedness to the goods, knowledge in sorting, space for storage, geographic location of the charity shops and their proximity to the rag dealers, and the value-added options available to the rag dealers.

The majority of clothing cast-offs start at charity shops such as Goodwill Industries or Salvation Army where they are examined for wearability in terms of mildew, tears and stains. The charity stores keep about 25 per cent of the goods donated for sale in their retail stores. The rest is sold by the kilo to rag dealers (Hawley, 2009). Even though there has been increased interest in consumers' willingness to buy second-hand clothing, the abundance of used clothing in the marketplace allows for consumer preference for top condition. Discerning consumers of used clothing are more willing to buy second-hand clothing the more often they do so (Hiller-Connell, 2009).

At a time when fast fashion is at its peak, forward thinking eco-boutiques have also realized an increased interest in carefully picked collectable and vintage goods. With the vast amounts of poor-quality goods savvy eco-boutique entrepreneurs of used collectables and vintage goods understand the value of hand-beaded garments, 100 per cent cotton Levis and 'real' lace.

One of the largest US sorting houses is Mid-West Textile Co. in El Paso, Texas. This company sorts a semi-trailer load of post-consumer textiles each day adding up to more than 4,535,923.7kg a year (S. Hull, personal communication, 12 February 2000). From these hundreds of thousands of kilograms, only 1–2 per cent is considered of collectable or vintage quality (personal communication with Thom Haxton, Mid-West Textile Co., 2009).

Not only can vintage and collectables be found in eco-boutiques in New York's Lower East Side or London's Camden Market, but also on the internet where more and more sites feature vintage or one-of-a-kind used clothing finds. Many UK owners of vintage shops are members of TRAID (Textiles Recycling for Aid and Development), a charity organization that finances itself through the sale of quality second-hand clothing. Current fashion trends are reflected by a team of young designers who use and customize second-hand clothes for a chain of specialty vintage clothing stores in the UK. TRAID offerings include 'cheap chic and occasional designer surprise' (Ojumo, 2002; Packer, 2002). In the US, used clothing boutiques are members of the National Association of Resale and Thrift Shops (NARTS). This Chicago-based association, founded in 1984, has more than 1000 members and serves thrift, resale and consignment shops and promotes public education about the vintage shop industry.

Today, vintage fashions have become a 'highly commodified fashion alternative to wearing new designs' (Palmer and Clark, 2005, p197). Vintage fashions appear in fashion and lifestyle magazines, on the internet and on celebrities. For zealots of vintage apparel, shopping becomes an obsession for the 'deal', the 'find' or the 'hunt'. Vintage shoppers pursue originality, a discerning difference in a world where retail merchandise has very much started to appear similar. Others seek nostalgia or identifiers with a celebrity or another culture. For example, Oprah recently auctioned some of her personal shoes and handbags on eBay to raise funds and awareness for her charity (Mitchell, 2010).

149

USED CLOTHING EXPORTS

On many street corners in developing countries, racks of used clothes from developed countries are being sold. The US exports nearly US$1 billion annually in sales to Africa. One of its primary export countries is Uganda, where Ugandi women can purchase designer blouses for less than a dollar. Most rag dealers in developed countries sort used clothing for the various markets in developing countries. In other words, lightweight clothing and sturdy shoes would go to sub-Saharan Africa, and men's and children's clothing would go to India, but women's clothing would not because women in India still primarily wear traditional dress. The sorted clothing is compressed into 272–450kg bales, wrapped and warehoused until an order is received for export. Several things are considered when sorting used clothing: climate of the market, relationships between the exporters and importers, and trade laws for used apparel. Recent negotiations between the US Department of Commerce and East African countries (Uganda, Kenya and Tanzania) resulted in a 300 per cent import tax. For example, the Tanzanian Bureau of Standards, cited the following concerns:

- requirements of fumigation certificates;
- ban on used undergarments, socks, stockings and nightwear;
- requirement that bales should not exceed 50kg;
- a requirement of a health certificate to prove the country of origin is free of diseases;
- certification that the garments are used to protect fledgling apparel manufacturing industries;
- a sampling of the consignment (Smartasn, 2010).

TOP LEFT | Figure 4.1.3a Blankets being made from used sweaters for IKEA

BOTTOM LEFT | Figure 4.1.3b Warp beam of yarns made of used sweaters

Yet cheap Chinese clothing imports dumped in Africa pose as much threat to local textile and apparel industry as second-hand clothing from Europe and the US (personal communication, E. Stubin, 10 February 2005). Furthermore, the used clothing trade provides many small business opportunities throughout Africa.

In the past, clothing was not carefully sorted by particular brand or style before it was exported from the US to Africa. But more recently, Africans have become more discerning consumers, keen on both style and price, causing rag dealers to sort more carefully before bales are shipped overseas. Sorting more carefully can add value to the bale and demand a higher price, but also means higher processing costs.

Used Western clothing in Africa is referred to as mitumba or salaula (Tranberg-Hansen, 2000). Sorted by quality and type, there is high demand for shirts, trousers, suits, T-shirts, jackets and licensed athletic wear. Used men's suits are of particular value, especially compared to new suits, which in retail stores are often out of the price range of many local residents. Mitumba serves a useful role to fill a gap between those who can afford and those who cannot. Mitumba also provides important entrepreneurship opportunities for many of Africa's poor as they clean, repair and sell the clothing in trade throughout the continent.

REDESIGNING

In recent years, designers and micro-design companies have been engaged in restyling from used clothing. Creating something new from old by cutting up and restyling is the imaginative and resourceful genius behind companies such as Junky Styling and XULY. Bët (clothes made from post-consumer waste), Harricana (redesigned furs) and From Somewhere (clothes designed from pre-consumer waste). Because used clothing is non-standard, each design is unique and original; but at the same time this design technique is labour intensive, resulting in higher prices. Consumers interested in redesigned or repurposed garments are usually already emotionally connected to the garment either because of an environmental philosophy or emotive connection, so price is less of a barrier.

Redesigned or repurposed clothing often comes with layers of meaning; at the same time that the history of the garment is evident, so a contemporary identity exists. When second-hand garments go through transformation, the past life of the garment is reshaped into something relevant for today. Redesigning is not the 'easy way' to design, but the result is unique one-of-a-kind or limited run garments. As Palmer and Clark point out, today's second-hand clothes are no longer simply a passage of items from those who have to those who do not. Today's second-hand clothing may, indeed, enter a 'new and different phase where actual items of clothing are not simply being transformed for their use value but become fashionable commodities that contribute to a complex global trade' (Palmer and Clark, 2005. p4).

OTHER SORTED CATEGORIES

Clothes that are not sorted for second-hand exports, vintage markets or repurposed designs are used for other value-added markets including wipers, stuffing for mattresses and pet-beds, and feeder stock for reproduced yarns. Unfortunately, little interest exists from industrial or waste engineers to find additional uses for textile waste and almost no research funds for innovation are siphoned to creation of value-added products made from used clothing as a feeder stock, even though some progress has been made from companies that make blankets from old sweaters, insulation from old blue jeans and geo-bales from recycled carpet (GeoHay, 2010).

One company that diverts cotton from the landfills is Jimtex with its Eco2cotton™, a recycled cotton suitable for socks and other apparel knits and decorative home items. Eco2cotton™ comes in a respectable range of 70 colour blends targeted to the high-end knit market. The response has been tremendous. Yet, even with companies like Jimtex who are innovating in creative ways, throughout the Western world many warehouses remain filled with baled used clothes waiting for other value-added ideas.

THE CHALLENGE

The textile sector is a complex system that impacts many other systems: agricultural, social, economic, political, technical, to name a few (Dickerson, 1998). For change to occur in such a large, complex and interconnected industry, we have to realize that even small change can have impact. When giant companies like Wal-Mart make efforts, significant change starts to occur. In early 2007, when Wal-Mart retired their familiar blue vests, the fibre from the vests was diverted from landfill and a partnership was formed with Hallmark, Inc. to produce 10,000 specialty greeting cards that were distributed to US troops serving in Iraq and Afghanistan (personal communication with Kim Brandner, network captain for Wal-Mart's Textile Sustainable Value, August 2007). Wal-Mart is committed to other sustainability initiatives including the organic cotton programme for baby clothes, sheets and towels resulting in a commitment of 8–10 million tonnes of organic cotton over five years, assuring farmers that there will be a market for their crops (Gunther, 2006). Wal-Mart's CEO Lee Scott is committed to sustainable initiatives. The company created the Sustainability Index Team consisting of members from a cross-section of people from corporate headquarters, non-government agencies, suppliers, academics and researchers, and strategic sustainability partners to start identifying corporate waste to share ideas, set goals and monitor progress. Over time, they are committed to eliminating overpackaging, revamping logistics to save fuel and creating 'sustainable value networks'. Eventually Wal-Mart hopes to have Sustainability Index ratings on a wide range of products for consumer evaluation. Yet critics continue to focus on Wal-Mart's social issues not recognizing that simply because of Wal-Mart's behemoth size, a small change can make a huge impact on the global supply chain and ultimately the environment.

151

While evidence exists of an increased consciousness toward sustainability in the textile industry, the work that is being done remains focused on niche problems or segments of the industry. As Fletcher points out, current sustainability approaches are limiting in terms of bringing 'fundamental change to the way the textile sector works' (Fletcher, 2009, p369). She goes on to note that in order to become truly sustainable, we will need to move out of our current comfort zone and imagine a society where a sustainable textile industrial system becomes a relational industrial system that depends on healthy social, environmental and economic interactions (Fletcher, 2009, p374). Using Meadow's visionary systems theory, Fletcher explores ways in which the textile industry system can be impacted. Perhaps most critical to sustainable and systemic change is addressing the primary goal of profit. This would require a paradigm shift that balances social and environmental concerns with economic concerns (Fletcher, 2009, p378).

AN ALTERNATIVE APPROACH

Since World War II, the volume and variety of clothing produced and consumed worldwide has risen swiftly. Even people in populous developing nations such as China, India and Indonesia, as well as smaller ones such as Vietnam, Ghana or Guatemala are now drawn into the global apparel consumption market as never before. Many groups who for centuries wore traditional dress now consume T-shirts and jeans saving traditional dress for ceremonial purposes.

The expansion of scale and the increased velocity of the market system are factors in changes in production, in the division of labour and in consumption patterns. Every day an increasing number of retailers must stay competitive so they offer more and more cheap fashion goods at low prices that have been grown, sewn, packaged and shipped from places far away at phenomenal speeds.

As far as fashion offerings are concerned, the choice is widely varied and often 'on trend'. Indeed, particular retailers that offer 'fast fashion' produce such trendy offerings that consumers flock to get the latest. Yet, the aggregate effect of fast fashion has been, I fear, a tolerance for marginal quality and poor service. With each new shipment of fast fashion, styles are pushed to the consumer even if the style change is only minimal. Therefore there is the argument that fast fashion will eventually destroy true style, resulting in a global baseline of fashion mediocrity.

At the same time, consumers have lost their ability to claim their own sense of style – following, instead, what the media or style-makers and trend-makers claim is style. Seldom does a consumer claim a sense of style that holds true to self over time, heeding instead what fashion dictates. If consumers had the ethical courage to claim personal style, then there would be less need for trendy fast fashion that had to be based on quick stock-turns, narrow profit margins, distant low-wage production and poor quality.

An alternative to fast fashion is slow fashion, modelled after Carols Petrini's concept of slow food. Clark's (2008) work on slow fashion begins the discussion, but Lehew and Hawley (2010) argue that slow fashion must work in tandem with mass manufactured fashion because of the difficult challenge of clothing the planet's population in a cost-effective manner. Slow fashion is a concept that could reduce the amount of excess consumption by honouring local and authentic, high quality and taste education. One way to initiate slow fashion is to develop personal style.

153

Claiming personal style would allow the return of strategic wardrobe purchases based on quality rather than quantity. Price would no longer be as important because purchases would be based on style rather than trend. Fashion could be assumed based on sustained notions of embracing something more relevant than 'planned obsolescence'. Imagine if what was truly trendy and embraced was the fact that you were embracing your personal style. To that end, you purchased one fabulous new item this year that may have cost as much as you have spent in recent years, but instead of ten poor-quality pieces, you have one fabulous blouse – or dress – or skirt, made by a local cooperative or of natural dyes, employing local workers – and it coordinates and builds on your existing wardrobe.

Companies such as Alabama Chanin or Mack & Mack in Greensboro, North Carolina provide examples of slow fashion. These are companies that have made philosophical choices to produce quality goods, with authentic and local producers. The result is fashion worth supporting because you know where it came from; therefore it has far more to do with your own personal style than the dictated trend of the moment.

Fast fashion will not go away. But perhaps over time, more and more people will search for their personal style as they embrace more fashions that are produced locally, authentically, and with consideration for quality and the environment. We will consider establishing and embracing our own sense of style rather than consuming what fast fashion dictates, only to throw it away for the next fast fashion that is dictated to us.

Fashion and its meanings are increasingly mobile in a globalized clothing market. But the same clothing has very different meanings in different contexts. Contemplate these juxtapositions: second-hand T-shirts with holes in a poverty-stricken rural area or a second-hand T-shirt screen-

LEFT | Figure 4.1.4 Junky Styling. VD Mac. Photographer Ness Sherry

printed with Grateful Dead concert dates and signed by Jerry Garcia; old moth-eaten wool sweaters that will be garneted back to fibres to be made into new blankets or a used cashmere sweater re-embellished and found in an eco-boutique in SoHo; old denim jeans salvaged to be made into housing insulation or vintage Levis bringing top dollar on the Paris auction block. These binaries force us to consider the contrasts of what happens to our used clothing. On the one hand, used clothing can seem vulgar and crude, but on the other hand, it offers economic, environmental and social encouragement for our future.

REFERENCES

Brosdahl, D. J. C. (2007) 'Quality over quantity: The key to sustainable fashion?' *Future Fashion. White Papers*. Earth Pledge, New York

Clark, H. (2008) 'Slow + fashion – an oxymoron – or a promise for the future...?', *Fashion Theory*, vol 12, no 4, pp427–447

Dickerson, K. (1998) *Textiles and Apparel in the Global Economy* (3rd edition). Prentice Hall, New York

Fletcher, K. (2009) 'Systems change for sustainability in textiles', in R. S. Blackburn (ed.), *Sustainable Textiles: Life Cycle and Environmental Impact*. Woodhead Publishing, CRC Press, Boca Raton, pp369–389

GeoHay (2010) www.geohay.com/ (accessed 25 April 2010)

Gregson, N. and Crewe, L. (2003) *Second Hand Cultures*. Berg, New York

Gunther, M. (2006) 'The green machine', *Fortune*, vol 154, no 3, pp42–48

Hawley, J. (2006) 'Digging for diamonds: A conceptual framework for understanding recycled textiles', *Clothing and Textiles Research Journal*, vol 24, no 3, pp262–275

Hawley, J. (2008) 'Economic impact of textile and clothing recycling', in J. Hethorn and C. Ulasewicz (eds), *Sustainable Fashion: Why Now?* Fairchild Books, New York, pp207–232

Hawley, J. (2009) 'Understanding and improving textile recycling: A systems perspective', in R. S. Blackburn (ed.) *Sustainable Textiles: Life Cycle and Environmental Impact*. Woodhead Publishing, CRC Press, Boca Raton, pp179–199

Hiller-Connell, K. Y. (2009) 'Exploration of second-hand apparel acquisition behaviors and barriers', *Proceedings of the International Textile and Apparel Association Conference in Seattle, WA*, November 2009. www.itaaonline.org/downloads/CB-Hiller-Connell-Exploration per cent20_of per cent20Second_Hand.pdf

Lee, M. (2007) 'Fast fashion', *The Ecologist*, vol 37, no 2, p60

Lehew, M. and Hawley, J. (2010) 'Slow fashion: Can we use the slow food movement as a model?', *Proceedings of the International Textile and Apparel Association conference in Montreal, Canada*, October 2010

Mitchell, C. (2010) 'Oprah sells personal items on eBay to help raise money for The Oprah Winfrey Leadership Academy'. www.examiner.com/x-21937-Kansas-City-eBay-Examiner-y2010m3d1-Oprah-sells-personal-items-on-eBay-to-help-raise-money-for-The-Oprah-Winfrey-Leadership-Academy (accessed 3 April 2010)

Ojumo, A. (2002) 'Charity shops are beating the high street at its own game', *The Observer*, 24 November, p57

Packard, V. (1963) *The Waste Makers*. David McKay Company, New York

Packer, G. (2002) 'How Susie Bayer's T-Shirt ended up on Yusef Mama's back', *New York Times*, 31 March, p54

Palmer, A. and Clark, H. (2005) *Old Clothes, New Looks: Second Hand Fashion*. Berg, Oxford

Smartasn (2010) *Secondary Materials and Recycled Textiles*. http:www.smartasn.org/news.html (accessed 26 March 2010)

SMART (2004) 'The textile recycling industry', *Canadian Textile Journal*, vol 121, no 6, pp30–35

Tranberg-Hansen, K. (2000) *Salaula: The World of Second Hand Clothing and Zambia*. University of Chicago Press, Chicago, IL

US EPA (Environmental Protection Agency) (2008) *Recycled Textiles*. www.epa.gov/wastes/conserve/materials/textiles.htm (accessed 30 May 2010)

Wilson, E. (2003) *Adorned in Dreams*. Rutgers University Press, Chapel Hill, NC

ABOVE | Figure 4.2.1 The clothing donation bank of The Smith Family organization in Sydney

4.2 CASE STUDY
REUSE IN THE FASHION CYCLE

Thriving businesses and non-profit organizations have textile waste at the core of their operations. The 'rag trade' has a long history of textile waste recycling, and as Strasser (1999) notes, waste is created by sorting and classifying. Through re-sorting and reclassifying, this waste can create opportunities for organizations to benefit and profit from. The Smith Family is an Australian children's charity that is partly supported by The Smith Family Commercial Enterprise, established in 1963. A part of this, The Smith Family Nonwoven Manufacturing Plant is located in Villawood, New South Wales and has been producing non-woven textiles since 1987. The Smith Family operates with both pre-consumer and post-consumer waste. Pre-consumer waste includes fabric clippings from the fashion and textile industries as well as donated unsold end-of-season clothing from fashion companies, while post-consumer waste consists primarily of donated second-hand clothes.

The Smith Family sorting facility, the largest such operation in New South Wales, has over 60 people sorting 10,000 tonnes of clothing onto conveyor belts each year. The clothing is sorted into five categories: retail, export, wiper, non-woven and waste. The Smith Family has to pay for any waste placed in donation bins to be landfilled. Winter clothes and Australian labels tend be sold in The Smith Family retail stores in Australia while most summer clothes are exported. Approximately 30 per cent of inventory in the retail stores is new, unworn clothes donated by fashion companies while 70 per cent is second-hand. Interestingly, an increase in new clothes tends to result in loss of customers. Between five and ten containers are exported each week. These quantities reflect a fashion-conscious, consumerist society. For clothing this may be the ideal form of recycling as not only is the clothing recycled in a materials sense but its purpose – to be worn – is also retained.

Donated clothes deemed unsuitable for retail or export may be sorted for wiper and non-woven textile production. Wipers are cut from suitable cloth, and cloth for non-woven textiles is baled and taken to a rag-tearing machine. Carding, needle punching and thermal bonding are used to manufacture a wide range of non-woven textiles. Alongside recycled fibres virgin fibres are used. The Smith Family also receives fashion industry fabric clippings for this purpose. Annually the company manufactures 400 tonnes of non-woven textiles from both new and recycled fibre. The non-woven consists of three lines: textiles for insulation, felt for agricultural, underlay and acoustic purposes. and textiles from virgin fibres for cleaning cloths, filtration products and wall panelling.

Design for disassembly is one strategy that could make certain operations easier for organizations such as The Smith Family. As some rag-tearing machines cannot cope with zips and buttons, making these durable for wear but easy to remove at end-of-life would facilitate repurposing garments that are no longer wearable. Furthermore, such organizations might benefit from collaborations with fashion designers or fashion students to refashion wearable but dated garments.

....

Strasser, S. (1999) *Waste and Want. A Social History of Trash*. Metropolitan Books/Henry Holt and Company, New York

ABOVE | Figure 4.2.2 The recycling process of collecting, sorting and distribution at The Smith Family organization in Sydney

4.3 CASE STUDY
A NEW FUTURE FASHION INDUSTRY

Fashion clothing contributes to millions of tonnes of landfill waste each year globally, which is a frightening reflection of the wastefulness created by the fashion system. As fashion designers begin to adopt sustainable strategies within their working practices, often through the use of organic or renewable materials, the quest to change the disposability of fashion remains unchallenged. It has been suggested that in some instances true disposability may be a positive thing if product materials flow through a closed loop system. To use McDonough and Braungart's (2002, p104) idea of material as 'nutrient', the Wonderland project opens a discussion about future possibilities.

The Wonderland project was a collaborative experiment that began as a conversation between designer and artist, Professor Helen Storey (London College of Fashion) and Professor Tony Ryan (Sheffield University), a polymer chemist. Exploring the notion of intelligent and sustainable packaging, the project investigated the concept of a disappearing water bottle and a water purification device. The bottle was created from a polymer than can be dissolved in hot water, which when cooled forms a gel material that can be used as fertilizer. Providing an ideal growing environment for herbs and plants, the material becomes a true nutrient. The intricate Wonderland dresses, developed from textiles created by Trish Belford (Interface, University of Ulster) emerged as a metaphor for our disappearing planet. Storey and Ryan state: 'This is our call to creative arms. We are exploring the power of shared ideas. We all have a duty to use our talents, our imaginations and our rigour to create a healthier planet.' But how might disposability fit in the context of sustainability in fashion design? For specific textile applications disposability is desirable and even legally required, for example in medical use. Typically, synthetic non-woven fibres are used; however, these can take centuries to decompose in landfill sites. While issues of potential contamination would need resolving, the polymer developed within the Wonderland project suggests an exciting direction.

If certain sectors of the fashion industry are going to continue to pursue fast fashion, then could a nutrient-based soluble polymer be a solution? The vast majority of fast-fashion garments are made from polyester and many of these are typically disposed of in landfill sites. However, the possibility of fertilizing a garden or vegetable patch with an unwanted garment would seem a safe alternative. Kate Fletcher and Mathilda Tham explored this scenario with the 'One Night Wonder' disposable party top in the Lifetimes project, which acknowledged that fashion exists across a range of speeds and rhythms (Fletcher, 2008, p176). The Wonderland project demonstrates that disposability could be an integral part of a sustainable future for fashion. No single solution will resolve the vast quantities of waste associated with the fashion industry, but Wonderland paves the way towards some of those solutions.

Fletcher, K. (2008) *Sustainable Fashion and Textiles*. Earthscan, London

McDonough, W. and Braungart, M. (2002) *Cradle to Cradle: Remaking the Way We Make Things*. North Point Press, New York

LEFT | Figure 4.3.1 Wonderland at London College of Fashion. Photographer Alex Maguire

OVER | Figure 4.3.2 Detail, Wonderland spider flowers. Photographer Alex Maguire

KATE FLETCHER

4.4 POST-GROWTH FASHION AND THE CRAFT OF USERS

INTRODUCTION

In order to bring about change, sustainability values and experiences have to be real to people. Yet we know very little about people's everyday encounters with fashion and sustainability. For instance, how does it feel to wear a garment that connects us with others? What is our understanding of engaging with nature through the clothes we wear? Just as each of these personal experiences is different, so is each person experiencing them. Yet most of our experience of fashion has no space for this heterogeneity: instead what we face is a limited set of fashion encounters, reproduced in Paris, London, New York or Tokyo. This is but one of many flashpoints that signal the profound nature of the sustainability challenge for the fashion sector. This is a challenge that requires us to transform not only fashion products and manufacturing processes but also fashion's context, its rules and goals, business models and methods of promotion. In this chapter I explore the interconnections between real, live experiences of sustainability in fashion and what they suggest about the shape and structure of the industry that creates the fashion and clothes that we wear. My interest is in amassing a collection of ideas that improve people's experience of fashion qualitatively without necessarily growing the industry in quantitative scale. Together these ideas offer a tentative glimpse of a new prosperity in fashion that exists outside the predominant economic and business model of growth that is so closely associated with fashion today.

THREE TALES OF USER INVENTIVENESS

It seems only appropriate that any exploration of the connections between the way people use clothes and sustainability should be grounded in real experience. Thus this chapter begins with three stories recorded as part of the Local Wisdom project that convey some of the garment-wearing ingenuity of the British public.

Edward: I call this my three stage jacket. It began about forty years ago as a very slim waistcoat that was given to me. I knitted a panel and put it into the back just to be able to fasten it together at the front, you see. And then about fifteen years ago I added sleeves and a collar and some trimmings. And then, only about five years ago, I became a bit too big to button it up so I added latchets across to the front so that I can fasten it.

Yvonne: This is a dress that I've had for 25 years and share with my sister. We sort of have it for 5 years each and then post it back to each other and it's like fancy dress for me... almost like cross dressing... it brings out a different part of me. At the moment I just wear it for special occasions but I once met a woman who was in her 80s and who wore eveningwear all the time. She'd made a decision years before not to buy any new clothes and to wear everything until it wore out. She'd worn her way through her wardrobe and had got to her eveningwear. So when I'm in my 80s I'm going to wear this dress...

Andy: In 1978 my Mum gave me £10 to buy a jacket and jeans and this is the one I bought. Back then I was a punk and I sewed badges on the back... Sex Pistols, Sham 69, The Stranglers... and my Grandad's RAF stripes on the arm. I've still never washed it... why would I? And anyway it would wreck the badges.

These three tales of alternative patterns of use offer one starting point for understanding more deeply the potential of the actions of users in shaping the sustainability of the fashion industry. Make no mistake, such individual stories of resourcefulness, thrift, emotional connection and social defiance are far from earth shattering in nature. They are rarely dramatic, instead they impose a human scale and intimacy on the insight they afford. Yet, for me, their small size is key. For actions like making a change to a seam or never laundering a garment that is worn time and again are eminently do-able activities and within the reach and influence of us all.

166

INTRODUCTION TO THE LOCAL WISDOM PROJECT

In 2009, I started the ongoing Local Wisdom project of which the tales above form a part, with the aim of recognizing and honouring culturally embedded sustainability activities in fashion that exist at the level of the user. It involves gathering images and stories from the general public about the way clothes are used with the hope of developing fresh understanding about more resourceful and satisfying use of garments. The project's premise is that sustainability can emerge from a wealth of simple interactions and in fashion it has potential to flow not only from a garment's design and production supply chain, but also from the choices we all make as users on a daily basis: how we select, wear, care for and connect with our garments. As such, opportunities to transform the sustainability potential of clothes are widely distributed throughout the population and not just reserved for designers, production managers or the fashion cognoscenti. They happen with each new insight, use and with every service of needle and thread in homes in far-flung corners of every nation; often far away from catwalks or business agendas.

The process of recording these widely distributed acts of inventiveness is very simple: a photo shoot is set up in various locations, in the first instance at three places in the UK. It is then widely advertised in the vicinity; signs are put up in newsagents' windows, in local libraries and sports centres. Advertisements are placed in local newspapers and interviews given on community radio networks. We network with local groups and so far have affiliated with Stitch and Bitch clubs, regional textile festivals, the climate change campaigning group Cape Farewell and the Transition Towns movement. We extend an open invitation to the public to attend the shoot with garments that fit into specific categories (more on this below). We then record the telling of the garment's

RIGHT (TOP) | Figure 4.4.1 Edward. Photographer Fiona Bailey

RIGHT (BOTTOM) | Figure 4.4.2 Yvonne. Photographer Fiona Bailey

ABOVE | Figure 4.4.3 Andy. Photographer Fiona Bailey

'story' in audio and photograph the volunteer participant in his/her piece. The emerging body of information is ad hoc, specific and often surprising. It is, by turns, interrogated and supported anew by the actual practices of clothes wearing captured at each additional Local Wisdom event.

While the call to participate in the Local Wisdom project is wide open, the categories of garments that people are invited to bring are more structured. They are designed specifically to tease out sustainability-supporting user-related activities, as distinct to producer-related ones. That is, to uncover the ingenuity and improvisation that goes on with and to clothes after the point of purchase. These are not necessarily done within the rubric of intellectualized concerns or commercial opportunities for sustainability, but instead emerge from the culturally embedded 'wisdoms' of thrift, domestic provisioning, care of community, freedom of creative expression and connectedness to nature, among other things. This explicit emphasis on the widespread practices of use, rather than the challenges of production, as a starting point for change towards sustainability signals a departure from what has gone before. Most work around sustainability themes in fashion to date is firmly focused on the manufacturing supply chain and lessening the (very considerable) impacts of agricultural practices, fibre mills, dye houses and cut-make-trim factories among others. Yet vital as it is, this work forms but a part of the sustainability challenge for the fashion sector. For what goes on *after* production processes are over and the garment has been sold – that is, the personal, variable, myriad use patterns that occur in homes and wardrobes – is also a key factor influencing sustainability in fashion, yet is often overlooked.

Local Wisdom offers a glimpse of these experience-based extant practices that, by design or default, also support sustainability objectives. Many of these practices are motivated by reasons other than environmental or social improvement, by memory or thrift, for example, or by a sense of common identity derived from wearing a garment that is owned by someone we care about. They reflect an array of ways of being, having and doing with clothes and provide an opportunity to celebrate these garment-related actions as valuable in their own right. They also provide us with an opportunity to explore them in a sustainability context. To ask, for example, if they reveal new insight conducive to enhanced well-being; if they remind us of traditional knowledge; or perhaps if they show how things can be done more efficiently. Local Wisdom is an exercise in empiricism, in gathering practical experience and ideas of many users in order to seed understanding about what type and form change towards sustainability may take in fashion, when the root of this change is the users of clothes, not their producers.

The garment categories drawn up for the Local Wisdom photo shoots are informed by some of the key learning that has been made in sustainability issues in fashion over the last two decades. For a whole host of structural, historical and financial reasons (Fletcher, 2008) most of this learning has been geared towards production and supply chain-related issues rather than practical understanding or cultural significance of garment use; nevertheless, there are a number of areas where the role and cultural understanding of users is widely acknowledged as critical in affecting a garment's sustainability. One such area is clothes' laundering, which has been shown to account for around 80 per cent of the lifecycle energy consumption of frequently washed garments (Franklin Associates, 1993). Another is the multifaceted area of garment durability and the complex emotional and psychological issues associated with making a garment last, rather than just making a long-lasting garment, and which have the potential to profoundly affect patterns of consumption and disposal (Chapman, 2005). Other categories pick up on familiar sustainability themes such as localism and 'lightness' (i.e. efficient use or distribution of materials) (Fletcher, 2008). Other more speculative categories are also included, based less on established data sets and energy calculations and more on a vision for how we might live in a sustainability-directed future. Eminent industrial ecologist John Ehrenfeld (Ehrenfeld, 2008), has suggested, for example, that products that foster a sense of connectedness with the natural world and with other humans are important in promoting sustainability – and the Local Wisdom process provides an opportunity to see whether people already use garments in this way. The nine categories used in the Local Wisdom project to date are:

- garments that are shared between people;
- garments that have never been washed – and aren't leather;
- garments that have the character of a particular place in them;
- garments that link you with the natural world;
- garments that catch your attention each time you wear them;
- garments that tell the story of how they've been used;
- garments that are made up of separate pieces that can be interchanged;
- garments that make you feel part of a community (but not a uniform);
- garments that are enjoying a third, fourth or fifth life.

THE WISDOM OF USERS

The culturally embedded practices revealed by the Local Wisdom process are, to date, largely made up of a set of pragmatic, materially frugal actions, often motivated by emotional triggers and frequently exemplified by old or second-hand (rather than new) clothes. This 'wisdom', in my view, adds quality and fine distinction to understanding the ways in which certain resourceful and satisfying practices work on the ground. For example, from the wide range of never-washed garments brought to the events, it became clear that a key influence in determining whether a piece can defy social pressure and never be laundered, is fear that the washing process itself causes something precious to be lost: a scent, a memory, the particular way a garment fits, the quality of handwork and even a political stance. This evocation of emotion as a major influence in home laundering practices stands at odds even with leading industry approaches, which treat laundering as a technical and behavioural function of wash cycle efficiency but not an emotional one.

Other wisdom reveals nuanced insight into what motivates people to share garments (mainly, it appears, to reinforce connecting bonds with others and to forge new shared experiences) and how people manage this process practically (posting a garment back and forth through the mail every few years; telephoning around a family group to see 'who has the dress'; etc.). Other wisdom still uncloaks the very great extent some people will go to in order to rework garments (and resources more generally) to meet their changing needs and express creativity and the associated expertise, sense of pride and satisfaction this brings. Indeed much of the clothes-based ingenuity gathered in this project so far appears to be a combination of practical technique and emotional skill; that is, head and hand, jointly employed to negotiate the symbolic rules and roles fashion and clothes play in people's lives.

For any reader it will, I am sure, be clear that such culturally embedded practices offer a set of vastly different starting points for change towards sustainability than those adopted by industry to date, for they privilege sensitivity to people's lived experience rather than industrial or commercial ideas about what sustainability is or should be. Not only are such practices personal, variable and slow to enact; falling outside of (mass) commerce, and hence fashion as we know it (which instead prefers standardized, global products that are quick to produce); they also fall outside many people's views about the intellectual scope of the sustainability challenge for industry and the 'proper' response to it (where it is often framed as a production-related issue to be solved by industry, technology and savvy resource management techniques). Yet it is my view that these sorts of user-initiated, culturally embedded practices hold potential to transform fashion sustainability in a new way. They offer an expanded view of the reality of sustainability practice on the ground; one that exists outside the boundaries within which designers, manufacturers and retailers currently operate. In addition, they sketch out the possible shape of a new layer or type of fashion commerce based on broader values than profit and sales growth, geared instead towards increasing the quality of fashion experience rather than its quantitative scale.

EXPANDING THE FASHION INDUSTRY'S SUSTAINABILITY FRAMEWORK

In the last two decades the intellectual framework that has most shaped sustainability work in the fashion industry (as in most other sectors) is lifecycle thinking. Simply, it involves directing attention not only to fibre type, material provenance or the processing stage that a company deals with directly, but also to the whole physical lifecycle of the product, from raw materials to end of life. It sees garments as a mosaic of interconnected flows of materials, labour and as potential satisfiers of needs and not simply as isolated resources, processes or sources of one-off environmental, social and cultural impact in production. Lifecycle thinking is inspired by the language and study of ecology and works to understand the countless interrelationships that link material, industrial and economic systems with nature. These connections operate at different scales and with different spheres of influence, some on a direct local level and others globally. In lifecycle thinking, openness to these relationships is a key precursor for change as it demonstrates the dynamic effect of each part on every other. The goal of this 'global' view is to optimize the sustainability of the whole product 'system' and for this to take a priority over increasing the effectiveness of individual system parts or lifecycle phases. And unlike in the past, where each lifecycle phase was the concern of discreet companies or industries (spinners, weavers, white goods manufacturers, etc.), now the success of the whole entails joint responsibility for all players to reduce impact. This makes things like waste, for example, as much a concern and focus for designers and consumers as for recyclers. Lifecycle thinking confers a distinctive way of framing sustainability problems as interconnected issues extending beyond the boundaries of individual companies or even industries (Heiskanen, 2002), so as to promote awareness, responsibility and optimal solutions from all those involved with designing, producing, using and reusing clothes.

By and large the practical 'on the ground' implementation of this intellectual framework is very far from the conceptual ideal. Evidence for this is widespread. In policy, for example, the European Commission's Integrated Pollution Prevention and Control regime sets out 'best available techniques' for the textile sector by chiefly focusing on wet processing stages and associated chemical use, therefore seeming to emphasize a partial improvement agenda rather than a fully integrated, whole lifecycle-based legal framework. In manufacturing, the highly fragmented structure of the textile processing chain, typically involving a large number of small and medium sized companies, results in a tendency for individual companies to work to bring about change to processes that chiefly bring benefit to themselves rather than the whole. In the sector more generally, the common preference for technology-based solutions to sustainability problems overlooks other sorts of innovation. Favouring technological fixes is perhaps inevitable in an industry like textiles that since the Industrial Revolution in the 18th century has been processing materials better, faster and cheaper by improving technology. However, the result is a tendency to neglect the substantial effect that users and other non-technologists have on determining a product's overall sustainability potential right across the lifecycle. It can also be seen that this technology bias overlooks the power and agency that culturally embedded practices like low energy use, garment refashioning and novel ways of clothes' wearing have in influencing sustainability. These practices, which reflect the reality of sustainability practice on the ground, exist outside the boundaries within which designers, manufacturers and retailers currently operate. For these designers, producers and high street stores tend to work in ways that are familiar, in areas where they have most control and where they will feel the benefits directly. They focus on materials and their provenance, production practices and logistics efficiency. What happens with users falls outside of this. Yet for me, the stories and images of culturally embedded 'wisdoms' associated with use are the essential companion to fashion design and production. For designing and using form a single whole: the one shapes the other. The process of feeding back user innovations and improvisations to designers inescapably influences the evolution of fashion practice over time and space. In honouring lived experience of the practices surrounding garment use, the broad and connective intellectual framework of lifecycle thinking is affirmed. New stakeholders (potentially all users) are brought into the process. Different ways of knowing, such as through experience or intuition, gain equal privilege to scientific rationalism.

Small acts of user creativity, resourcefulness, emotional significance or defiance have been called 'a user's craft' and are described by philosopher Richard Sennett as 'live intelligence fallibly attuned to the actual circumstances of life' (Sennet, 2008). This craft, displayed unassumingly by many of the Local Wisdom project participants, deals with the metamorphosing of the form, application and way of using material objects, like garments, over time. It holds a mirror up to the multiple interconnections between people, resources and products and shows the potential of experience on the ground and in wardrobes to influence the sustainability of a garment's design. Very practically, it also provides us with an array of starting points, ideas and pragmatic

171

examples of a more satisfying use of fashion resources; though one that is studded with radically different expressions of material status, ways of behaving, emotional connections and power relations to the established norm. Using it as the basis of practice turns the design process out on itself, changing its goals and ideas. John Ehrenfeld sees such change of process and thought as absolutely necessary for sustainability: 'Achieving positive results requires drastic action. We need to shift from our reductionist, problem-solving mode to one that is driven by a vision of a sustainable future we all share. We need to reflect carefully on our current state of affairs and replace ineffective ways of thinking and acting' (Ehrenfeld, 2004).

DESCRIBING A NEW TYPE OF COMMERCE

In fostering sustainability through effective thought and action, and capturing expressions of this (as in the Local Wisdom project), a set of changed economic opportunities begins to emerge. This contrasts sharply with the priorities of today's fashion industry, which is structurally reliant on economic growth tied to expanding resource use: on making and selling increasingly more units to improve market share, increase profit and stay in business. Here the economy grows in physical scale (see Figure 4.4.4) and because the planetary ecosystem in which the economy sits is of fixed size; relative to it, the economy grows continually larger. The growth imperative that shapes daily decisions in fashion businesses (like the vast majority of others) is fundamentally at odds with the finite nature of the resource base and fragile ecosystems upon which we depend for survival. In the last 60 years the size of the global economy has increased by a factor of five (Jackson, 2009) and the default assumption is that this will go on expanding indefinitely in both poor countries, where better quality of life is unquestionably needed, and rich nations, where it has been shown that material wealth – the goal of economic growth – adds little to happiness. At the same time, a slew of indicators reveal the implications of this economic structure on environmental and social quality: compared to just two generations ago, poverty is just as endemic, with 2 billion people still living on less than US$2 a day; social cohesion, particularly in the rich West, is weaker (Hamilton, 2003); atmospheric carbon concentrations are at far higher levels; natural environments are more degraded; and there are growing numbers of conflicts over land use and access to water (Stern, 2007).

For an increasingly vocal body of commentators, the great contradiction implicit in promoting growth based on a continually expanding scale of resource use, 'as the cure for all economic and social ills' (Daly, 1992), is motivating the formulation of alternative economic structures and social practices designed to foster prosperity without growth (Jackson, 2009). The goal is to disassociate material throughput from commercial success; and to define and describe economic activity by ecological limits. One of the forerunners of this 'post-growth' economics is Herman Daly, who 30 years ago set out ideas for a balanced and bounded 'steady state economy' (Daly, 1992). In Daly's steady state economy, there is a constant stock of physical capital that is capable of being maintained by a low rate of material throughput that is within the regenerative and assimilative capacities of ecosystems (see Figure 4.4.5). Daly defines a steady state economy in physical terms – the resource creating and pollution-absorbing limits an ecosystem places on the economy – not as zero-growth economic activity. In a steady state economy, commerce is alive and well, just operating in different places and layers; 'the end of physical accretion is not the end of progress. It is more a precondition for future progress, in the sense of qualitative improvement' (Daly, 1992, p182).

It goes without saying that the building of an economic framework that cultivates qualitative improvement without growth poses a profound challenge for the fashion sector. It raises numerous questions not least, for example, what 'quality' means beyond resource use and how that influences fashion's output and role, which today is a fusion of both material and message. It queries what

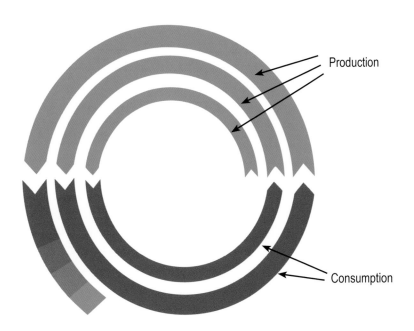

ABOVE | Figure 4.4.4 Ever-growing cycles of production and consumption (Daly, 1992). Such a view can encourage an economy that can ultimately strain the environment

173

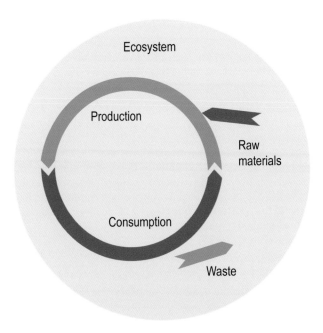

ABOVE | Figure 4.4.5 Steady-state economics considers cycles of production and consumption that take the surrounding ecosystem into account and try to achieve equilibrium with it (Daly, 1992)

the physical limits of the fashion sector are, if bounded by a healthy ecosystem. It also asks about the scale of 'throughput' (i.e. consumption of new garments and disposal of old ones) that the ecosystem is able to regenerate and assimilate safely. While accurate answers to these questions are still to be developed, it is clear from ecological and social evidence as varied as climate change, growing waste mountains and persistent global poverty, that the physical scale of today's (fashion) economic system is unsustainable. While the numbers are still to be worked out, the truth is stark: in a post-growth economy, the fashion industry's trade in physical product would shrink dramatically from its current levels.

In an economic system geared to the optimum scale of total resource use relative to the ecosystem, rather than to growth, the quantity of physical fashion product is held at a steady level. What is not held steady and is free to expand is knowledge, creativity, ingenuity, the success of our relationships, the quality of our experiences and how satisfied we are. Cultural capital can increase. Wealth can be redistributed and resources allocated differently. A post-growth economy is an economic system that 'develops qualitatively but not in quantitative scale' (Daly, 1992, p182). Its effect, according to Tim Jackson in his report for the UK Government (Jackson, 2009, p34), is to provide us with 'bounded capabilities' that help us prosper and live well within clearly defined limits.

It is into this context of 'bounded capabilities' and flourishing within limits that culturally embedded practices of garment use seem to find a natural home. For the craft of users needs little money, materials or physical capital to make it happen: it does not expand resource use. Instead it works to improve the real, live experience of using garments through the application of expertise, ingenuity and freethinking. The craft of users often results in garments being kept in service for longer, perhaps through repair and refashioning (such as in the case of Edward's three stage jacket mentioned in the introduction), or through the forming of powerful emotional connections (as in the case of Yvonne's shared dress). The effect is to delay disposal and, if keeping them in active service prevents a replacement being bought, reduce the throughput of physical goods through the economy. Other tales of ingenuity in use, such as Andy's never laundered jacket, influence the same agenda but through a different route. Rather than slowing the churn of garments through the economy, they improve the resourcefulness of each item as it is used.

More than that however, the stories and images of user inventiveness captured as part of Local Wisdom seem to infer an increase in *quality*. By this I don't mean better quality materials or more expert making techniques (though this too is possible), but rather an increase in the quality of engagement and connection that people have with their garments and, because of the self-improvement nature of much of this activity, also with themselves as human beings. This improved connection has been called 'true materialism' and contrasts with the sort of materialism prevalent today. Sociologist Juliet Schor cites the cultural critic Raymond Williams when she says, 'we are not truly materialist because we fail to invest deep or sacred meanings in material goods. Instead our materialism connotes an unbounded desire to acquire, followed by a throwaway mentality' (Schor, 2002).

To be in a state of engagement and connection, people have to be active and able, to have access to skills, tools and opportunities to use them. They have to be recast in roles other than just as consumers but also as competent individuals and suppliers of ideas and skills to the fashion system. Much has been written about the psychological and sustainability benefits of shifting away from 'a life of consumption' or one dominated by 'having', to what Ivan Illich calls 'a life of action' (Illich, 1975) or in Erich Fromm's terms, life in the 'being' mode of existence (Fromm, 1976). In a small way the practices unearthed in Local Wisdom can be seen to be part of this shift and convey a seizing of initiative by users of clothes to actively change or improvise their patterns of use. As such they express one way to improve quality of fashion experience within the physical limits of the clothes we already have. They reflect, albeit tentatively, one set of activities that comprise fashion in a post-growth economy. They bring those wisdoms borne of using clothes as part of everyday life into direct contact with the future sustainability of the fashion industry. And with that they bring the prospect of trading cultural capital and users' knowledge in a new layer of exchange or commerce that is measured in terms broader than growth and increasing use of resources and instead based on increasing the quality of the fashion experience.

REFERENCES

Chapman, J. (2005) *Emotionally Durable Design: Objects, Experiences and Empathy*. Earthscan, London

Daly, H. (1992) *Steady-State Economics* (2nd edition) Earthscan, London, pp180, 182

Ehrenfeld, J. (2004) 'Searching for sustainability: No quick fix, *Reflections*, vol 5, no 8, p4

Ehrenfeld, J. (2008) *Sustainability by Design*. Yale University Press, New Haven, CT

Fletcher, K. (2008) *Sustainable Fashion and Textiles: Design Journeys*. Earthscan, London

Franklin Associates (1993) *Resource and Environmental Profile Analysis of a Manufactured Apparel Product: Woman's Knit Polyester Blouse*. American Fiber Manufacturers Association, Washington, DC

Fromm, E. (1976) *To Have or To Be*. HarperCollins, London

Hamilton, C. (2003) *Growth Fetish*, Pluto Press, London, p34

Heiskanen, E. (2002) 'The institutional logic of lifecycle thinking', *Journal of Cleaner Production*, vol 10, pp427–437

Illich, I. (1975) *Tools for Conviviality*. Fontana, London

Jackson, T. (2009) *Prosperity without Growth*. Earthscan, London

Schor, J. (2002) 'Cleaning the closet', in J. B. Schor and B. Taylor (eds), *Sustainable Planet*. Beacon Press, Boston, p55

Sennet, R. (2008) *The Craftsman*. Penguin Books, London, p199

Stern, N. (2007) *The Stern Review: Economics of Climate Change*. http://webarchive.nationalarchives.gov.uk/+/www.hm-treasury.gov.uk/media/3/6/Chapter_1_The_Science_of_Climate_Change.pdf (accessed 31 March 2010)

CASE STUDY DESIGNER PROFILES

UPCYCLING MATERIALS FOR FASHION: Romance Was Born

Australian design duo Anna Plunkett and Luke Sales created the label Romance Was Born five years ago after becoming friends studying fashion together at East Sydney Technical College. Their technical ability and intricate attention to detail have enabled them to develop a truly unique brand with an unmistakable aesthetic. Like creative bowerbirds they assemble a fashion mosaic drawing together elements of all that is beautiful to create their stunning embellished style. Their witty, imaginative style and handcrafted aesthetic have found them a stellar following, with Cate Blanchett naming Romance Was Born as one of her favourite labels, and rock royalty Deborah Harry, Cyndi Lauper, Yeah Yeah Yeahs, Bat for Lashes and M.I.A. among those to sport their one-off ensembles on international stages.

Further information: **www.romancewasborn.com**

NEW MATERIALS FOR FASHION: Jennifer Shellard

Jennifer holds a fractional academic research post at the London College of Fashion and runs the BA Fashion Jewellery course. She studied an MA in Constructed Textiles at the Royal College of Art having earlier completed a first degree in three-dimensional design at Manchester Polytechnic. Her practice-based research has included two AHRC awards, culminating in the recent Light Cloth project, which explores colour perception through a combination of handwoven textiles, time, light and colour. She has taken part in residencies and professional practice in the UK, Latvia and Sri Lanka and has exhibited and lectured widely within the UK and abroad.

Further information: **www.fashion.arts.ac.uk/jennifer-shellard.htm**

177

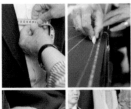

THE TAILOR'S CRAFT: Bijan Sheikhlary

Bijan began his tailoring career at the age of 13 in Iran. He was taught the craft of handmade tailoring in the bespoke tradition and by the time he was 17 acquired the title of master jacket maker. At 21 Bijan was acknowledged as a master tailor and was considered one of the finest bespoke tailors in Iran. His continual striving for perfection led him to Saville Row. It was here at H. Huntsman and Sons, considered by many to be the finest bespoke tailors in Savile Row, that Bijan honed his craft and continued to evolve his art of tailoring. To Bijan, bespoke tailoring is not just a craft but an art. He considers himself to be an artist whose medium is clothing the human form.

Further information: **www.bespokebijan.com.au**

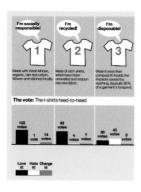

THE PERFECT SOLUTION: Better thinking Ltd

Better thinking gives you a fresh perspective on your problems – and a 21st century way of solving them. Clients work directly with Mark and Mike, the two strategic and creative founding partners, and benefit immediately from their global expertise in strategy, design, sustainability and engagement and their unique insight into the key disciplines that matter to businesses today: user experience, digital, mobile, brand, social media and corporate social responsibility. They are unconventional – and driven by strong principles – creative, rigorous, open, inclusive and collaborative. These qualities, combined with a methodology that is always 100 per cent focused on the project objectives, is how they deliver better thinking in every project.

The more complex and systemic the problem is, the more satisfying it is for them to present their clients with a simple and powerful solution. They thrive on solving unconventional and complex problems... the ones that stubbornly persist as thorns in the side of any business. They enjoy getting under the skin of a brand or business issue, and shaping a powerful solution to the problem. And unlike some traditional consultancies, they are just as keen and able to tackle 'unglamorous' problems as those that are in the media spotlight.

Further information: **www.betterthinking.co.uk**

SLOWING FASHION: Gene Sherman

Dr Gene Sherman AM is Chairman and Executive Director of Sherman Contemporary Art Foundation. She was formerly Director and Proprietor of Sherman Galleries, representing major artists across Australia and the Asia-Pacific region (1986–2007). She is a current Board member of the National Portrait Gallery, Art & Australia magazine and the Australia–Israel Cultural Exchange. Dr Sherman's awards include the Chevalier de l'Ordre des Arts et des Lettres (2003), Doctorate of Letters honoris causa (University of Sydney, 2008) and Member of the Order of Australia (2010).

Further information: **www.sherman-scaf.org.au**
                     www.powerhousemuseum.com

PERSONALIZING FASHION: Alex Martin

Alex is a choreographer, performer, designer and seamstress in projects that defy easy categorization. Alex lives in Columbia City, Seattle, Washington, US with her partner Freya, their son Ari, William the cat, a small but bountiful garden and a flock of super-cute chickens. Alex is better at turning to the left than to the right, loves learning new things and she likes to build things that last.

Further information: **www.littlebrowndress.com**

REUSE IN THE FASHION CYCLE: The Smith Family

The Smith Family is a national, independent children's charity that works in partnership with other caring Australians to help disadvantaged Australian children. The organization's work is informed by research that enables The Smith Family to work with communities effectively to build their capacity to improve outcomes for children and young people and their families. The Smith Family Commercial Enterprise supports this work in the community by generating untied funds and offsetting the costs of running the organization. This allows The Smith Family to put more of the donations directly into programmes rather than administration costs. For over 40 years, The Smith Family has been supplying high-grade quality products to customers all around the world.

Further information: **www.thesmithfamily.com.au**

A NEW FUTURE FASHION INDUSTRY: Helen Storey

Professor Helen Storey is an artist and designer living and working in London. After graduating from Kingston Polytechnic in 1981, Storey worked in Italy and London before launching her own label in 1983. Having won Most Innovative Designer of The Year in 1991 and nominated for British Designer of The Year by the British Fashion Council, Storey established the Helen Storey Foundation with Caroline Coates in 1997. A not-for-profit arts organization that promotes creativity and innovation, the Helen Storey Foundation has led collaborative projects across multiple disciplines. In 2008, Wonderland, a project that straddles the junction between art, fashion and science, was developed with Professor Tony Ryan at Sheffield University and reached an audience of 11 million people. Storey was awarded Honorary Professorships at Heriot Watt University and King's College London in 2001 and 2003 respectively and was awarded visiting Professor of Chemistry at Sheffield University in 2008. Storey is a Senior Research Fellow and Professor of Fashion and Science at the London College of Fashion and Co-Director of the Fashion Science Centre. She was awarded the MBE for services to the Arts in June 2009.

Further information: **www.helenstoreyfoundation.org**
www.fashion.arts.ac.uk/5192.htm
www.wonderland-sheffield.co.uk
www.wonderland-belfast.co.uk
www.fashion.arts.ac.uk

RESOURCES
READING LISTS

Fashion and textiles

Black, S. (2008) *Eco-chic: The Fashion Paradox.* Black Dog Publishing, London

Blackburn, R. S. (ed.) (2005) *Biodegradable and Sustainable Fibres.* Woodhead Publishing, Cambridge, UK

Blackburn, R. S. (ed.) (2009) *Sustainable Textiles: Life Cycle and Environmental Impact.* Woodhead Publishing, Cambridge, UK

Blanchard, T. (2007) *Green is the New Black.* Hodder & Stoughton, London

Brown, S. (2010) *Eco Fashion.* Lawrence King, London

Dickson, M.A., Loker, S. and Eckman, M. (2009) *Social Responsibility in the Global Apparel Industry,* Fairchild Books, US

Earthpledge (2008) *FutureFashion White Papers.* Earthpledge, New York

Fletcher, K. (2008) *Sustainable Fashion & Textiles: Design Journeys.* Earthscan, London

Hethorn, J. and Ulasewicz, C. (eds) (2008) *Sustainable Fashion: Why Now? A Conversation Exploring Issues, Practices, and Possibilities.* Fairchild Books, New York

Lee, M. (2007) *Eco Chic: The Savvy Shoppers Guide to Ethical Fashion.* Gaia Books Ltd, London

Oakes, S. R. (2008) *Style, Naturally: The Global Guide to Sustainable Fashion and Beauty.* Chronicle Books, San Francisco

Rivoli, P. (2005) *The Travels of a T-Shirt in the Global Economy: An Economist Examines the Markets, Power, and Politics of World Trade.* John Wiley & Sons, Hoboken, NJ

Ross, A. (ed.) (1997) *No Sweat: Fashion, Free Trade and the Rights of Garment Workers.* Verso Books, New York and London

General design

Birkeland, J. (2002) *Design for Sustainability.* Earthscan, London

Brower, C. and Mallory, R. (2005) *Experimental EcoDesign: Architecture/Fashion/Product.* RotoVision, Hove

Chapman, J. (2005) *Emotionally Durable Design.* Earthscan, London

Chapman, J. and Gant, N. (2007) *Designers, Visionaries and Other Stories.* Earthscan, London

Lewis, H. and Gertsakis, J. (2001) *Design and Environment: A Global Guide to Designing Greener Goods.* Greenleaf Publishing, Sheffield, UK

McDonough, W. and Braungart, M. (2002) *Cradle to Cradle: Remaking the Way We Make Things.* North Point Press, New York

Schor, J. and Taylor, B. (eds) (2002) *Sustainable Planet: Solutions for the Twenty-first Century.* Beacon Press, Boston

Thorpe, A. (2007) *The Designer's Atlas of Sustainability.* Island Press, Washington, Covelo and London

Walker, S. (2006) *Sustainable by Design: Explorations in Theory and Practice.* Earthscan, London

WEB RESOURCES

Researchers and projects

Alex Martin/Little Brown Dress: www.littlebrowndress.com

Ann Thorpe http://designactivism.net

Becky Earley www.beckyearley.com

Centre for Sustainable Fashion www.sustainable-fashion.com

Fashioning Now www.fashioningnow.com

Kate Fletcher www.katefletcher.com

Kate Goldsworthy www.kategoldsworthy.co.uk

Kathleen Fasanella www.fashion-incubator.com

Local Wisdom www.localwisdom.info

Teach Sustainability www.teachsustainability.com.au

TED – Textile Environment Design www.tedresearch.net

The Uniform Project www.theuniformproject.com

Organizations

Ethical Clothing Australia
www.ethicalclothingaustralia.org.au

Ethical Fashion Forum
www.ethicalfashionforum.com

Fairwear Australia www.fairwear.org.au

Made-By www.made-by.nl/?lg=en

Sourcemap www.sourcemap.org

Sustainable Style Foundation
www.sustainablestyle.org

The Clothing Exchange Australia www.
clothingexchange.com.au

The Environmental Justice Foundation
www.ejfoundation.org

The New Economics Foundation
www.neweconomics.org

The Smith Family http://thesmithfamily.com.au

Fabrics

All Eco www.alleco.com.au

Aurora Silk www.aurorasilk.com

Certton www.certton.com.au

Eco Source www.hempweave.com

Greenfibres www.greenfibres.com

Hemp Gallery www.hempgallery.com.au

Hemp Wholesale Australia www.hempwa.com

i-Merino www.i-merino.com

Instyle Textiles www.instyle.com.au

KimoYes www.kimoyes.com

Loop Fabric www.loopfabric.co.uk

Macquarie Textiles
www.macquarietextiles.com.au

NearSea Naturals www.nearseanaturals.com

Organic Cotton Directory
www.organiccottondirectory.net

Silk Road Fabrics www.srfabrics.com

Standardknit www.standarduniversal.com.au

Vreseis www.vreseis.com

Fashion labels

Alabama Chanin http://alabamachanin.com

Bird Textiles www.birdtextile.com

EDUN www.edunonline.com

Elsom www.elsom.com.au

Flora 2 www.flora2.com

From Somewhere www.fromsomewhere.co.uk

Globe Hope www.globehope.com

Junky Styling www.junkystyling.co.uk

Katherine Hamnett http://katherinehamnett.com

Keep & Share www.keepandshare.co.uk

Linda Loudermilk http://lindaloudermilk.com

Loomstate www.loomstate.org

Mark Liu http://markliu.co.uk

Marks & Spencer Plan A: http://plana.
marksandspencer.com

Materialbyproduct www.materialbyproduct.com

Patagonia www.patagonia.com

Perfect t-shirt / Luxury Redefined
www.luxuryredefined.co.uk

Rachael Cassar http://rachaelcassar.com

ROGAN NYC www.rogannyc.com

Romance Was Born http://romancewasborn.com

Study NY http://4equalsides.com

Online magazines

Ecotextile News www.ecotextile.com

Ecouterre www.ecouterre.com

Inhabitat www.inhabitat.com

Style Will Save Us: www.stylewillsaveus.com

The Ecologist www.theecologist.org

TreeHugger www.treehugger.com

181

EDITORS' BIOGRAPHIES

ALISON GWILT is a fashion and textiles academic, researcher and curator who has been exploring sustainable strategies for fashion and textile design practice through her practice-based research activities and PhD studies since 2003. Alison gained a degree in fashion and textiles at the Central St Martins College of Art in London, and her career in academia includes managing courses and lecturing in fashion and textiles design in the UK, New Zealand and Australia, where she is currently at the University of Technology Sydney (UTS). Alison has presented papers at international conferences, contributed to publications and press, and has been the recipient of competitive funding from organizations such as the New South Wales Government's Environmental Trust for which she was Chief Investigator for the project, *Fashioning Now: Changing the Way We Make and Use Clothes*.

TIMO RISSANEN is Assistant Professor of Fashion Design and Sustainability at Parsons The New School for Design in New York. He is a fashion-design academic and fashion designer whose design practice is informed by inquisitive pattern-making and sustainability concerns. His PhD project with strong practical focus, titled Fashion Creation Without Fabric Waste Creation, examines an aspect of fashion design in which the fashion designer and pattern maker can directly effect positive change. Rissanen previously taught fashion design at the University of Technology Sydney (UTS), Australia. He has presented at several international conferences and he contributed a chapter to *Sustainable Fashion. Why Now?* (Fairchild Books, 2008).

INDEX